Industrial Engineering Step by Step

Fundamentals of Improving Operational Processes and Complex Systems

Robert Thomas

© 2024 by Robert Thomas

All rights reserved.

No part of this publication may be reproduced, distributed, or transmitted in any form or by any means, including photocopying, recording, or other electronic or mechanical methods, without the prior written permission of the publisher, except in the case of brief quotations embodied in critical reviews and certain other noncommercial uses permitted by U.S. copyright law.

This book is intended to provide general information on the subjects covered and is presented with the understanding that the author and publisher are not providing professional advice or services. While every effort has been made to ensure the accuracy and completeness of the information contained herein, neither the author nor the publisher guarantees such accuracy or completeness, nor shall they be responsible for any errors or omissions or for the results obtained from the use of such information. The contents of this book are provided "as is" and without warranties of any kind, either express or implied.

PREFACE

Welcome to *Industrial Engineering Step by Step*. This book is designed as a guide for anyone interested in understanding, applying, or advancing within the field of industrial engineering. Whether you're a student, a professional looking to expand your knowledge, or simply curious about how complex systems operate more effectively, this book is for you.

Industrial engineering (IE) might sound intimidating at first. It's a field that spans the design, improvement, and implementation of integrated systems—systems that often include people, machines, materials, energy, and information. But when we break down IE into simple steps, you'll see it's all about making things work better. As we go through the chapters, I hope to introduce you to a field that's both powerful in its impact and accessible in its application.

This book covers the foundational concepts, frameworks, and techniques that industrial engineers use to solve problems and optimize processes. We begin with an introduction to the field, exploring its evolution, the important role of industrial engineers, and the principles of efficiency and productivity that form the backbone of IE. Each chapter is structured to give you a clear overview of specific topics, practical insights, and relevant techniques you can apply in various industries.

In *Chapter 1: Introduction to Industrial Engineering*, you'll get an overview of IE, including its history, purpose, and the industries where it's most impactful. This chapter will set the foundation for everything we cover throughout the book. Understanding IE as a discipline—its ethical principles, challenges, and connections to other fields like management and operations—will help you see where industrial engineers fit within larger systems and projects.

After this introduction, we go into the practical side of industrial engineering. Chapters 2 through 16 cover critical concepts that build on each other. *Chapter 2: Process Analysis and Design* discusses mapping processes and finding bottlenecks, which is essential for creating streamlined workflows. From there, *Chapter 3* introduces work measurement and time studies, key for assessing efficiency.

We'll move on to more advanced topics like facility layout in *Chapter 4*, lean manufacturing in *Chapter 5*, and Six Sigma and quality control in *Chapter 6*. These chapters explore how to improve physical spaces and manage production techniques to cut down on waste, improve quality, and deliver better outcomes. Throughout these chapters, I've included strategies to make IE principles applicable to real-world scenarios, even if you're new to the field.

Chapter 7, on supply chain management, tackles logistics, demand forecasting, and inventory management, covering how industrial engineers help manage the flow of goods and services from suppliers to customers. Following this, *Chapter 8* goes into operations research and optimization techniques—mathematical frameworks used to make the best decisions possible when resources are limited.

Another important aspect of IE is ergonomics, which we explore in *Chapter 9*. Ergonomics focuses on human factors, from physical design to health and safety, ensuring that workplaces are both efficient and comfortable for people. In *Chapter 10*, we address project management—skills that are essential for planning, scheduling, and managing resources effectively, especially in complex engineering projects.

Inventory management, cost analysis, and systems engineering are covered in Chapters 11 through 13. Each of these areas provides insights for controlling costs, analyzing decisions, and integrating systems, allowing you to understand how industrial engineers contribute to organizations' financial and operational success. Then, *Chapter 14* introduces data analytics, emphasizing the role of data in IE, from predictive analytics to machine learning. You'll see how data-driven decision-making is changing the field.

Chapter 15 covers simulation and modeling, tools that allow engineers to test scenarios before implementing them, which is especially helpful in high-stakes environments. And in *Chapter 16*, we look at sustainability and green engineering—a growing area within IE. Here, you'll learn about environmental impact assessments, energy efficiency, and the principles of a circular economy, focusing on making industrial practices more sustainable.

Finally, *Chapter 17: Other Tidbits* wraps up with practical advice for those considering a career in IE, along with key terms and definitions that you may find useful.

Throughout the book, each concept is explained step-by-step, with real-world examples and insights to make learning industrial engineering accessible and engaging. My goal is to give you a thorough understanding of IE principles and practices, without overwhelming you with technical jargon or overly complex explanations. You'll find that industrial engineering is a dynamic field that's as much about creativity and problem-solving as it is about technical know-how.

As you read, I encourage you to think about how these ideas apply to everyday situations. Industrial engineering isn't just for large-scale manufacturing plants or complex supply chains—it can be useful in virtually any field, from healthcare to technology, even in your personal life. The principles you'll learn in this book can help you think more critically about how things work and how they could work better.

I hope *Industrial Engineering Step by Step* inspires you to see the world with a new perspective and motivates you to make processes, systems, and environments more efficient and effective. Welcome to the field of industrial engineering. Let's get started.

TOPICAL OUTLINE

Chapter 1: Introduction to Industrial Engineering
- Definition and Scope of Industrial Engineering
- Historical Evolution of Industrial Engineering
- The Role of Industrial Engineers in Today's Workforce
- Fundamental Principles of Efficiency and Productivity
- Interdisciplinary Connections: Engineering, Management, and Operations
- Common Challenges in Industrial Engineering
- Overview of Key Concepts, Frameworks, and Techniques
- Ethics and Professional Responsibilities in Industrial Engineering
- Emerging Trends and Future Directions in the Field
- Industries and Applications: Where IE is Most Impactful

Chapter 2: Process Analysis and Design
- Mapping and Documenting Processes
- Identifying Bottlenecks and Inefficiencies
- Redesigning Processes for Optimal Flow
- Analyzing Process Cycle Times and Lead Times

Chapter 3: Work Measurement and Time Studies
- Quantifying Work for Efficiency
- Time Study Techniques
- Standard Data and Predetermined Motion Time Systems

Chapter 4: Facility Layout and Material Handling
- Principles of Effective Facility Layout
- Types of Layouts: Product, Process, and Cellular
- Material Handling Techniques
- Integrating Automation in Material Flow

Chapter 5: Lean Manufacturing Principles
- Understanding Lean and its Core Concepts
- Identifying and Reducing Waste (Muda)
- Techniques for Just-in-Time (JIT) Production
- Value Stream Mapping in Lean Systems

Chapter 6: Six Sigma and Quality Control

- Fundamentals of Six Sigma
- Defining and Measuring Quality
- Process Control and Statistical Quality Control (SQC)
- Root Cause Analysis and Problem-Solving Tools
- Implementing Control Charts for Process Monitoring

Chapter 7: Supply Chain Management

- Key Components of Supply Chains
- Demand Forecasting and Inventory Control
- Supplier and Customer Relationship Management
- Logistics and Distribution Planning

Chapter 8: Operations Research and Optimization Techniques

- Linear Programming and Optimization Models
- Decision Trees and Decision-Making Tools
- Simulation Modeling in Process Optimization

Chapter 9: Ergonomics and Human Factors

- Understanding Human Capabilities and Limitations
- Workstation Design and Ergonomic Assessment
- Cognitive Ergonomics and Usability
- Health and Safety in the Workplace

Chapter 10: Project Management for Industrial Engineers

- Project Life Cycle and Phases
- Planning and Scheduling with Gantt Charts and CPM/PERT
- Resource Allocation and Budgeting
- Risk Assessment and Mitigation

Chapter 11: Inventory Management and Control

- Types of Inventory and Inventory Costs
- Economic Order Quantity (EOQ) Models
- Just-in-Time (JIT) and ABC Analysis
- Inventory Tracking and Management Systems
- Safety Stock Calculations and Reorder Points

Chapter 12: Cost Analysis and Financial Decision-Making

- Fundamentals of Cost Analysis
- Cost-Benefit Analysis and ROI
- Breakeven Analysis and Profit Planning
- Financial Metrics for Industrial Engineers

Chapter 13: Systems Engineering and Integration

- Principles of Systems Thinking and Systems Engineering
- Systems Integration in Complex Operations
- Managing Interdependencies and Interfaces
- Cyber-Physical Systems and Industry 4.0

Chapter 14: Data Analytics and Industrial Engineering

- Basics of Data Collection and Analysis
- Statistical Methods for Process Improvement
- Predictive Analytics in Industrial Engineering
- Data Visualization and Interpretation
- Applying Machine Learning Techniques for Process Insights

Chapter 15: Simulation and Modeling in Industrial Engineering

- Types of Simulation Models and Applications
- Discrete-Event Simulation and Process Modeling
- Analyzing Simulation Data for Decision Making
- Implementing Simulation for Process Improvement

Chapter 16: Sustainability and Green Engineering

- Principles of Sustainable Industrial Engineering
- Environmental Impact Assessment
- Energy and Resource Efficiency Strategies
- Designing for Environmental Compliance and Resilience
- Circular Economy Principles in Industrial Design

Chapter 17: Other Tidbits

- How to Become an Industrial Engineer
- Terms and Definitions

Afterword

TABLE OF CONTENTS

Chapter 1: Introduction to Industrial Engineering ... 1
Chapter 2: Process Analysis and Design ... 18
Chapter 3: Work Measurement and Time Studies ... 25
Chapter 4: Facility Layout and Material Handling ... 30
Chapter 5: Lean Manufacturing Principles ... 37
Chapter 6: Six Sigma and Quality Control ... 44
Chapter 7: Supply Chain Management ... 52
Chapter 8: Operations Research and Optimization Techniques ... 60
Chapter 9: Ergonomics and Human Factors ... 67
Chapter 10: Project Management for Industrial Engineers ... 73
Chapter 11: Inventory Management and Control ... 79
Chapter 12: Cost Analysis and Financial Decision-Making ... 88
Chapter 13: Systems Engineering and Integration ... 94
Chapter 14: Data Analytics and Industrial Engineering ... 101
Chapter 15: Simulation and Modeling in Industrial Engineering ... 110
Chapter 16: Sustainability and Green Engineering ... 118
Chapter 17: Other Tidbits ... 125
Afterword ... 130

CHAPTER 1: INTRODUCTION TO INDUSTRIAL ENGINEERING

Definition and Scope of Industrial Engineering

Industrial Engineering (IE) is about designing, improving, and optimizing systems to make processes more efficient, cost-effective, and productive. At its core, it's the engineering discipline focused on understanding how to better organize complex systems involving people, machinery, materials, energy, and data. This field is unique in its interdisciplinary scope, blending concepts from engineering, management, and social sciences to address real-world problems across industries.

The **definition** of industrial engineering is often explained as "the design, improvement, and installation of integrated systems of people, materials, information, equipment, and energy." While many engineering disciplines focus on technical and mechanical aspects, industrial engineering is especially focused on creating efficient systems that maximize productivity with minimum waste. This includes not only tangible resources like raw materials and machinery but also intangible factors such as time, labor, and energy.

To understand the **scope of industrial engineering**, let's look at the types of systems industrial engineers work with and what they aim to improve. Unlike other engineering branches that may specialize in a single component, industrial engineering addresses systems holistically, viewing each part as interdependent. An industrial engineer's job is to identify where resources are not being used effectively, analyze the causes, and implement solutions. This may involve designing a production line to improve output, structuring workflows to minimize downtime, or reengineering supply chains to reduce costs.

Industrial engineers work in virtually every industry. In **manufacturing**, they design efficient production processes, ensuring minimal downtime and maximum product quality. For example, if a factory produces electronic components, the industrial engineer might analyze each stage of production to ensure materials and labor are utilized optimally. In **healthcare**, industrial engineers study patient flows, reduce wait times, and improve hospital operations. They may design scheduling systems for staff or reconfigure the layout of emergency rooms to reduce bottlenecks in patient care. In the **service industry**, they streamline operations, improve customer experience, and reduce operational costs. Here, they might work on optimizing queue times, inventory management, or staffing schedules.

A significant part of industrial engineering is **work measurement and productivity analysis**. By studying tasks, times, and motions, industrial engineers determine the most efficient ways to perform processes. They often use techniques like time studies and predetermined motion time systems (PMTS) to understand and standardize work activities. This kind of analysis is essential for establishing realistic work standards, improving labor productivity, and even determining fair wage rates based on workload.

Another area within the scope of IE is **quality control**. Industrial engineers analyze quality control data and implement systems like Six Sigma, a method that reduces defects and improves quality by identifying and eliminating variations in processes. By applying statistical tools, they can set benchmarks for quality, identify the root causes of defects, and develop strategies to maintain or improve product quality over time.

The **logistics and supply chain** component of industrial engineering ensures that materials and products flow smoothly from suppliers to customers. Industrial engineers work on optimizing inventory levels, choosing efficient transportation routes, and even setting up warehouse layouts. For instance, in a supply chain for a large retail company, they might design the system for ordering and storing goods so that products are consistently available to customers without excessive inventory buildup.

Industrial engineering also covers **facility layout and design**, where engineers determine how to arrange equipment, machinery, and workstations in physical spaces for optimal flow and accessibility. The goal is to minimize movement within facilities, reduce handling costs, and ensure that workflows are as smooth and efficient as possible. In a warehouse, this might mean placing high-demand items closer to shipping areas to reduce picking time.

In addition to physical processes, industrial engineers focus on **data analysis and optimization**. They use tools from operations research, including linear programming and simulation models, to solve complex decision-making problems. For example, a linear programming model might help determine the optimal production mix of products to maximize profit given certain constraints, like labor availability and raw material limits. Industrial engineers frequently rely on these mathematical models to guide decision-making, enabling managers to allocate resources more effectively and make informed choices under uncertainty.

Given the focus on all aspects of operational systems, industrial engineering provides the foundation for highly effective, data-driven decision-making that improves productivity, enhances quality, and ensures the most efficient use of resources. Through their unique expertise, industrial engineers can identify opportunities for improvement in nearly any process, saving companies time and money while improving overall performance.

Historical Evolution of Industrial Engineering

The history of industrial engineering is a journey through the development of productivity and efficiency in manufacturing and business operations. Its roots date back to the Industrial Revolution, where the demand for increased production volumes led to groundbreaking changes in how work was organized and managed. Early pioneers recognized that improving efficiency wasn't just about improving machines—it was about optimizing human and organizational factors as well.

Industrial engineering's evolution began in the 18th century with the rise of mechanized production. As factories replaced craft shops, **early efforts to improve efficiency** focused on machinery and labor organization. Managers sought ways to maximize output, but there was little scientific approach. This era set the stage for two of the earliest figures in the field: Frederick Winslow Taylor and Frank and Lillian Gilbreth.

In the late 19th and early 20th centuries, **Frederick Winslow Taylor** emerged as the "father of scientific management." Taylor applied a systematic approach to labor processes by breaking down tasks, timing each element, and establishing "one best way" to complete a job. His studies at the Midvale Steel Company, where he analyzed and improved metal cutting techniques, demonstrated how production could be dramatically increased through scientific analysis. This approach, often called "Taylorism," emphasized standardization, time studies, and clear task definitions. Taylor's principles led to higher productivity but were criticized for treating workers as mere extensions of machinery.

Around the same time, **Frank and Lillian Gilbreth** built on Taylor's work, focusing on motion study and ergonomics. They sought to reduce unnecessary movements and make work easier for employees. The Gilbreths' motion studies led to practical applications in tool design, workbench arrangement, and even surgical procedures. They are famous for analyzing the motions involved in tasks as common as bricklaying, creating methods that reduced fatigue and increased productivity. Lillian Gilbreth also introduced the human element into scientific management, recognizing the importance of worker welfare and contributing to the fields of industrial psychology and human factors engineering.

The early 20th century also saw the emergence of **Henry Gantt**, known for creating the Gantt chart, a tool still widely used in project management. Gantt's focus on scheduling and workflow visualization helped managers organize complex tasks and assess progress against deadlines. His work marked a shift toward a more holistic approach, considering not only tasks but also timelines and interdependencies.

The development of **World War II-era production methods** added new dimensions to industrial engineering. Faced with urgent demands for aircraft, vehicles, and weapons, manufacturers had to find ways to rapidly increase production while minimizing resource use. Techniques like assembly lines, mass production, and quality control evolved during this period. This led to the introduction of quality management principles, with engineers like **W. Edwards Deming** and **Joseph Juran** emphasizing statistical process control and the importance of continuous quality improvement. Their work laid the foundation for Six Sigma and modern quality management practices.

The post-war period saw **Japan's rise in manufacturing innovation**, significantly influencing industrial engineering. Japanese companies, notably Toyota, introduced methods like Just-in-Time (JIT) production and lean manufacturing. JIT minimized inventory by ensuring that materials arrived just as they were needed in the production process. Lean manufacturing, developed by **Taiichi Ohno** and **Shigeo Shingo**, emphasized waste reduction and continuous improvement (kaizen). The Toyota Production System became a model of efficiency and quality, inspiring global adoption of lean principles.

By the 1970s, the **rise of computer technology** transformed industrial engineering further, enabling engineers to analyze complex systems more effectively. With the advent of digital modeling, simulation, and data processing, industrial engineers could tackle problems at unprecedented scales. Operations research techniques, including linear programming and systems engineering, became integral to solving logistical and resource allocation problems.

In recent decades, industrial engineering has continued to evolve, incorporating data analytics, artificial intelligence, and sustainability principles. The **advent of Industry 4.0**, characterized by smart factories and interconnected systems, has introduced new challenges and opportunities for industrial engineers. Today, they apply digital tools to manage data flows, monitor real-time production, and optimize supply chains across global networks.

The evolution of industrial engineering reflects a progression from manual optimization to data-driven, integrated systems thinking. It's a discipline shaped by technological advancements and shifts in production philosophy, from Taylorism's strict task control to today's emphasis on human-centered design and environmental responsibility. Each phase has added new tools, techniques, and perspectives, building a rich legacy of continuous improvement and adaptability that defines the field today.

The Role of Industrial Engineers in Today's Workforce

Industrial engineers are integral to organizations worldwide, driving efficiency, improving processes, and reducing costs in diverse sectors. They bridge the gap between management and operations, applying scientific methods to design and optimize complex systems. In today's workforce, industrial engineers are highly valued not only for their technical skills but for their ability to bring measurable, strategic improvements across varied operational areas.

One of the most crucial roles of an industrial engineer is to **increase productivity**. This often involves analyzing workflows, identifying bottlenecks, and streamlining tasks to ensure resources are used optimally. For example, in manufacturing, an industrial engineer might observe assembly lines to see where delays occur and redesign the layout or sequence of operations to eliminate downtime. By doing so, IEs improve the speed and efficiency of processes without compromising quality.

In addition to productivity enhancement, industrial engineers focus on **quality control**. They use methodologies like Six Sigma and statistical process control to reduce defects and maintain high standards. Industrial engineers often collaborate with quality assurance teams to identify potential issues before they escalate, thereby preventing costly recalls or customer dissatisfaction. In industries like automotive manufacturing, where precision is vital, industrial engineers ensure products meet strict quality standards by refining both processes and inspection methods.

Supply chain management is another critical area where industrial engineers add value. With global supply chains becoming more complex, industrial engineers design systems that keep goods flowing efficiently from suppliers to customers. They use data to forecast demand, set inventory levels, and manage logistics, ensuring the right products are available at the right time while minimizing excess inventory. Industrial engineers work on optimizing these logistics networks, reducing transportation costs, and enhancing warehouse operations. In retail, for example, IEs can help manage inventory in a way that prevents stockouts without overstocking, achieving a balance that supports both cost-efficiency and customer satisfaction.

Industrial engineers are also responsible for **reducing waste and improving sustainability** within organizations. Applying lean manufacturing principles, they identify sources of waste in processes, whether it's excess material usage, time delays, or energy consumption. By minimizing waste, industrial engineers not only cut costs but contribute to environmental sustainability. Many organizations are now focused on reducing their carbon footprint, and IEs are essential in designing energy-efficient systems, sustainable production methods, and processes that minimize resource use.

In today's data-driven world, industrial engineers increasingly rely on **data analysis and digital tools** to solve complex problems. From simulation software to predictive analytics, these tools allow them to model scenarios, test solutions, and predict outcomes with greater accuracy. For example, in healthcare, industrial engineers use data analytics to improve patient scheduling and reduce wait times, leading to better patient care and resource utilization. The ability to interpret data and apply it in real-time decision-making has made industrial engineers invaluable in fields beyond traditional manufacturing, such as healthcare, logistics, and IT.

Industrial engineers serve as **problem-solvers and efficiency experts** across industries. They are important in adapting to emerging challenges, like supply chain disruptions or sustainability demands, by developing flexible, resilient processes. Their ability to work with cross-functional teams, manage complex projects, and optimize resources has made industrial engineers essential in creating adaptable systems that thrive in today's dynamic and competitive landscape.

Fundamental Principles of Efficiency and Productivity

Efficiency and productivity are the cornerstone goals of industrial engineering, and they involve maximizing output while minimizing resource use. To achieve this, industrial engineers apply a range of principles and methods that revolve around eliminating waste, optimizing processes, and using resources intelligently. By understanding and implementing these principles, organizations can achieve higher productivity, lower costs, and improved quality—results that directly impact profitability and competitiveness.

One key principle is **eliminating waste**, often referred to as the "seven wastes" in lean manufacturing: overproduction, waiting time, transportation, excess inventory, motion, over-processing, and defects. Waste elimination is central because it focuses on removing any activity that doesn't add value to the end product or service. For instance, if workers spend excessive time searching for tools, the industrial engineer's role is to reorganize the workspace to ensure items are within reach, reducing non-productive time and enhancing workflow.

Standardization is another vital concept. By standardizing tasks and processes, industrial engineers establish a consistent and repeatable method for each operation, which leads to predictable outcomes, reduced variation, and fewer errors. Standardization is often used in assembly line production, where each worker performs the same set of steps with minimal deviation. By creating and adhering to standardized procedures, industrial engineers ensure that the best practices are consistently applied across an organization, improving both productivity and quality.

Another fundamental principle is **work measurement**, which involves analyzing the time taken to perform a task. Industrial engineers use methods like time studies and work sampling to understand how long each process should take under ideal conditions. These measurements then serve as a baseline to identify areas for improvement, set realistic productivity targets, and establish fair labor standards. For instance, if data shows a particular task consistently takes longer than expected, industrial engineers investigate to find the root cause—whether it's due to inefficient workflow, outdated tools, or a lack of training—and implement corrective actions.

Ergonomics is also a central principle in enhancing efficiency and productivity. By designing workspaces and tasks that align with human capabilities and limitations, industrial engineers reduce physical strain on workers, minimize fatigue, and prevent injuries. For example, an ergonomically designed workstation for a manufacturing task might include adjustable seating, reachable tool placement, and automated lifting equipment to reduce the risk of repetitive strain injuries. Ergonomics contributes to efficiency by creating a safer and more comfortable working environment, which directly impacts productivity by reducing downtime and improving worker satisfaction.

The concept of **continuous improvement** underpins all efforts to enhance efficiency and productivity. Known in Japanese as "kaizen," continuous improvement encourages small, incremental changes to processes and practices rather than large-scale overhauls. Industrial engineers frequently adopt kaizen to identify and resolve minor inefficiencies. By focusing on continuous improvement, organizations develop a culture of proactive problem-solving where employees at all levels seek ways to refine and enhance their work.

Another foundational principle is **optimizing resource utilization**. Industrial engineers strive to use resources—including materials, labor, energy, and capital—efficiently to maximize output while keeping costs down. This often involves scheduling tools, personnel, and machinery to avoid idle time, overworking, or underutilization. For example, using forecasting models to predict demand, industrial engineers can adjust production schedules to prevent resource wastage and ensure that resources are allocated precisely where they're needed.

A closely related principle is **bottleneck management**, which involves identifying and alleviating constraints within a production system. A bottleneck is any step in the process where capacity is limited, resulting in delays and lower output. Industrial engineers analyze production flows to find these points and design solutions to increase capacity, such as reallocating resources, improving task efficiency, or adding equipment. Effective bottleneck management ensures that each part of the process flows smoothly, minimizing delays and optimizing productivity.

Lean manufacturing principles also are important in enhancing efficiency and productivity. Lean focuses on reducing waste, improving flow, and delivering value to the customer with fewer resources. Tools like value stream mapping help industrial engineers visualize the entire production process, pinpoint areas of waste, and make targeted improvements. By focusing on value-added activities and eliminating non-value-added ones, lean manufacturing promotes efficient resource use and higher productivity.

Lastly, **automation** is a principle industrial engineers apply to improve efficiency and productivity, especially in repetitive and time-intensive tasks. By incorporating automation technologies—such as robotic assembly lines or computer-controlled machinery—organizations can reduce human error, increase production speed, and maintain consistent quality. However, industrial engineers must carefully analyze which processes are suitable for automation, balancing costs and benefits to ensure that automation truly adds value to the organization.

Together, these principles of efficiency and productivity form a toolkit that industrial engineers apply across industries to design processes that are as effective, streamlined, and sustainable as possible. Through a methodical approach to eliminating waste, standardizing processes, and continually improving, industrial engineers are important in achieving optimal operational performance.

Interdisciplinary Connections: Engineering, Management, and Operations

Industrial engineering is inherently interdisciplinary, drawing from engineering, management, and operations to develop systems that improve productivity, reduce waste, and ensure efficient use of resources. The blend of these fields allows industrial engineers to address problems from multiple perspectives, creating solutions that are technically sound, managerially feasible, and operationally effective. This unique positioning within the workforce enables industrial engineers to serve as connectors and problem-solvers, bridging gaps between technical design, management strategy, and day-to-day operations.

At its core, industrial engineering is part of the larger **engineering discipline**, with foundations in mechanical, electrical, and systems engineering. Industrial engineers apply engineering principles—like mathematics, physics, and materials science—to solve real-world problems. For example, in designing a manufacturing process, they might draw on principles from mechanical engineering to select equipment, electrical engineering to design control systems, and systems engineering to ensure all parts of the process work together seamlessly. This technical foundation is essential for designing reliable systems and improving processes, but IE's focus extends beyond pure engineering into optimizing human and organizational factors as well.

Industrial engineering also incorporates key elements of **management**, especially in the areas of process improvement, quality control, and resource allocation. By applying management theories, industrial engineers help organizations make strategic decisions that improve productivity and profitability. For instance, they might use inventory management models to control stock levels, avoiding both stockouts and excess inventory costs. Additionally, industrial engineers often employ project management methodologies—such as Gantt charts and critical path method (CPM) scheduling—to manage and optimize large-scale projects. This management perspective allows them to align their technical solutions with the organization's overall objectives, creating value that resonates across departments.

Operations is another field that intersects deeply with industrial engineering. Operations management focuses on the efficient transformation of resources into goods or services, and industrial engineering provides the tools and techniques to refine these operations. For example, by applying lean manufacturing principles, industrial engineers streamline workflows, reduce waste, and increase efficiency in production. They might also use Six Sigma methodologies to ensure quality in every phase of the process, preventing defects that could disrupt operations. In service industries, industrial engineers work on optimizing service delivery, minimizing wait times, and ensuring customer satisfaction, applying operations principles to achieve seamless and efficient processes.

Industrial engineers often serve as **project coordinators or liaisons** between engineering teams and management, ensuring that technical aspects of projects are feasible within budgetary and operational constraints. This role requires industrial engineers to communicate complex engineering requirements to non-technical stakeholders while also translating management's strategic goals into actionable engineering solutions. For instance, in a manufacturing project, an industrial engineer may balance the technical requirements of production machinery with financial constraints, ensuring the project remains cost-effective without sacrificing quality.

Another interdisciplinary connection exists in **data analysis and information systems**. Industrial engineers rely on data to make informed decisions, which often involves analyzing operational metrics, financial data, and customer feedback. They might use data from information systems to monitor production, predict inventory needs, or improve scheduling. In doing so, they draw on knowledge from both engineering statistics and information technology, bridging gaps between technical data and management decision-making. Industrial engineers also use optimization algorithms from operations research, such as linear programming, to solve complex logistical problems, integrating their technical knowledge with strategic insights to find the best solutions.

Human factors and ergonomics is another field that connects industrial engineering with psychology and human sciences. By studying human behavior, capabilities, and limitations, industrial engineers design work environments that support productivity and minimize fatigue. In a warehouse, for instance, they might arrange workstations and tools to reduce strain on employees, while also considering operational flow. Understanding how humans interact with systems enables industrial engineers to create environments that are not only efficient but also safe and comfortable for workers, which in turn enhances overall productivity.

Lastly, **sustainability and environmental science** are increasingly integrated into industrial engineering. Industrial engineers are now tasked with creating processes that are not only efficient but also environmentally responsible. This involves understanding environmental regulations, developing energy-efficient systems, and designing for minimal waste generation. By applying engineering principles alongside environmental science, industrial engineers help organizations reduce their environmental impact and meet sustainability goals without sacrificing efficiency.

Industrial engineering's interdisciplinary nature enables it to address a broad range of challenges, combining engineering's technical rigor with the strategic focus of management and the practical approach of operations. This combination allows industrial engineers to create solutions that are both innovative and feasible, supporting organizations in achieving their goals in a way that is both efficient and adaptable.

Common Challenges in Industrial Engineering

Industrial engineers face a range of challenges, from balancing cost constraints to managing complex systems. The need to optimize efficiency while maintaining quality, safety, and adaptability in diverse settings is one of the primary difficulties in this field.

One common challenge is **managing resource constraints**. Industrial engineers are often tasked with improving processes or increasing production output without increasing budgets or resources. This requires them to think creatively and make trade-offs between time, cost, and quality. For example, when streamlining production lines in a manufacturing facility, industrial engineers may face limitations in terms of machinery, space, and labor. To address these constraints, they must prioritize improvements that yield the highest impact with minimal resources, often requiring careful analysis to find solutions that don't sacrifice quality.

Resistance to change is another major hurdle. Industrial engineers introduce changes that often disrupt established routines, impacting workers and management alike. Employees may worry that efficiency improvements could lead to job losses or changes in workload, while management may resist the upfront costs of implementing new systems. To overcome this, industrial engineers need strong communication skills and must demonstrate how proposed changes will benefit the organization and employees alike. For instance, they may need to provide data-backed evidence showing how a new process will not only save costs but also make employees' tasks easier and safer.

In modern industrial engineering, **data management and analysis** present significant challenges. With the advent of digital manufacturing and Industry 4.0, industrial engineers work with large volumes of data from various sources, including machine sensors, inventory systems, and customer feedback. Processing and analyzing this data accurately can be overwhelming without the right tools or expertise. Moreover, ensuring data accuracy is crucial, as poor data quality can lead to flawed decision-making. Industrial engineers must become proficient in data analytics tools and often work alongside IT departments to harness the potential of data-driven insights effectively.

Quality control and maintaining high standards in complex systems is another challenge, particularly in industries like automotive or aerospace, where precision is vital. Ensuring consistent quality while scaling production or introducing new technology requires careful planning and monitoring. Industrial engineers use techniques like Six Sigma and statistical process control (SPC) to monitor quality metrics, but implementing these systems across large-scale operations can be demanding. They must also troubleshoot issues quickly to prevent costly defects from reaching customers.

Another frequent challenge is **optimizing complex supply chains**. In today's global economy, supply chains are interconnected and often involve multiple stakeholders, including suppliers, manufacturers, and distributors spread across various regions. Industrial engineers must consider fluctuating costs, transportation logistics, and geopolitical risks when optimizing supply chains. They use tools like linear programming to design efficient routes and balance supply with demand, but external factors like shipping delays or trade restrictions can still impact the system. Being prepared to adapt to these uncertainties is essential for industrial engineers managing modern supply chains.

Lastly, **incorporating sustainability and environmental responsibility** is a growing challenge in industrial engineering. As more companies adopt sustainability initiatives, industrial engineers are expected to design processes that minimize waste, reduce energy consumption, and meet environmental regulations. This can be difficult to achieve without impacting efficiency or cost. Industrial engineers must innovate by integrating eco-friendly practices, such as using recyclable materials or designing energy-efficient systems, while still meeting operational targets. They often collaborate with environmental scientists and regulatory bodies to develop solutions that balance sustainability with productivity.

These challenges require industrial engineers to have a versatile skill set that combines technical expertise with adaptability, problem-solving, and interpersonal skills. Their ability to tackle these hurdles directly impacts an organization's success and its ability to compete in a fast-paced and resource-constrained world.

Overview of Key Concepts, Frameworks, and Techniques

Industrial engineering is built upon a foundation of core concepts, frameworks, and techniques that drive process efficiency, resource optimization, and productivity improvement. These methodologies equip industrial engineers to solve complex problems systematically, making operations smoother, cost-effective, and reliable.

A foundational concept in industrial engineering is **process mapping and analysis**, which involves documenting each step of a process to identify inefficiencies and improvement opportunities. Industrial engineers use tools like flowcharts and value stream mapping (VSM) to visualize the process and find areas where time, materials, or labor are wasted. For instance, in a production line, mapping out each stage helps engineers pinpoint bottlenecks, excess movement, or unnecessary steps. Process mapping not only identifies inefficiencies but also provides a clear structure for making improvements.

Lean manufacturing is a widely used framework aimed at maximizing value by minimizing waste. Originating from the Toyota Production System, lean principles focus on eliminating the "seven wastes" (overproduction, waiting, transportation, processing, inventory, motion, and defects). Techniques such as Just-in-Time (JIT) production ensure materials arrive precisely when needed, reducing inventory costs and improving workflow. Value stream mapping, an essential lean tool, helps industrial engineers identify and remove non-value-added activities, thereby streamlining production. Lean has become a universal standard in manufacturing and is now applied across various industries to achieve efficient operations.

Six Sigma is another critical methodology, focused on quality control and reducing process variation. Using data-driven approaches, Six Sigma aims to achieve nearly defect-free operations (3.4 defects per million opportunities). The Six Sigma framework is structured around the DMAIC process: Define, Measure, Analyze, Improve, and Control. This process helps industrial engineers systematically identify quality issues, develop solutions, and implement controls to sustain improvements. For example, in a pharmaceutical production setting, Six Sigma tools like root cause analysis and statistical process control (SPC) help prevent quality issues that could affect patient safety.

Time and motion studies are techniques industrial engineers use to understand and improve how tasks are performed. By observing and timing each movement in a task, engineers can standardize efficient work practices, eliminate unnecessary motions, and set realistic time standards. These studies contribute to work measurement, enabling fair labor standards and providing benchmarks for productivity. For example, in assembly line work, optimizing movements can reduce worker fatigue, prevent repetitive strain injuries, and boost output.

Operations research provides industrial engineers with mathematical tools to make optimal decisions, especially in resource-constrained environments. Techniques like linear programming, decision trees, and simulation allow engineers to model scenarios, evaluate options, and choose the most efficient solutions. For instance, in supply chain logistics, linear programming might be used to determine the optimal distribution route that minimizes transportation costs while meeting demand. Simulation modeling helps engineers test various scenarios virtually, giving insight into how changes might impact the system without disrupting real operations.

Ergonomics and human factors engineering focuses on designing systems that align with human capabilities and limitations. This approach ensures that work environments are safe, comfortable, and conducive to productivity. Ergonomics covers everything from workstation layout to tool design, and it's especially valuable in preventing injuries related to repetitive tasks. For example, in a warehouse setting, ergonomic principles might be applied to reduce the physical strain on workers who frequently lift or reach, improving both safety and efficiency.

Another crucial framework is **supply chain management**, which encompasses the planning, control, and optimization of the flow of goods from suppliers to customers. Industrial engineers use forecasting, inventory control, and logistics optimization to ensure goods are available at the right time and place without excess cost. Techniques like the Economic Order Quantity (EOQ) model help determine the ideal order size, balancing inventory costs with ordering costs. In addition, demand forecasting methods allow organizations to predict product demand and align production schedules accordingly.

Project management is essential in industrial engineering to ensure that projects stay on time and within budget. Tools like Gantt charts and the Critical Path Method (CPM) enable engineers to plan and monitor project schedules, track progress, and manage resources effectively. For example, in a new facility setup, industrial engineers use these techniques to coordinate tasks, ensure materials and equipment are available, and minimize delays.

Lastly, **data analytics** has become integral to modern industrial engineering. With advances in data collection and processing, industrial engineers analyze large datasets to gain insights into operations. They use predictive analytics to anticipate future demands, track performance metrics in real time, and adjust processes dynamically. For instance, data-driven analysis can reveal patterns in equipment maintenance needs, allowing engineers to implement preventive maintenance and reduce unexpected downtime.

Together, these concepts, frameworks, and techniques form a comprehensive toolkit that industrial engineers use to optimize operations across various industries. This interdisciplinary approach allows them to address complex challenges and drive continuous improvement, ensuring that processes are as efficient, productive, and sustainable as possible.

Ethics and Professional Responsibilities in Industrial Engineering

Ethics and professional responsibility are fundamental in industrial engineering, given the field's broad impact on productivity, human welfare, and the environment. Industrial engineers often work on systems that affect large numbers of people, including employees, consumers, and the public. With this influence comes a duty to ensure that all improvements or optimizations are not only efficient but also ethically sound and socially responsible.

A primary ethical consideration for industrial engineers is **safeguarding worker health and safety**. Designing efficient processes should never come at the expense of employee welfare. Engineers must adhere to safety standards and regulations, such as those from OSHA, to minimize workplace hazards. For example, in designing assembly lines or manufacturing layouts, industrial engineers must consider ergonomic factors to reduce strain, fatigue, and the risk of injury. Ignoring these aspects can lead to serious physical consequences for workers, undermining the ethical foundation of the profession.

Industrial engineers also bear the responsibility of ensuring **fair labor practices**. Since they often analyze and set work standards and productivity goals, they must avoid pushing workers to meet unrealistic quotas or overextending them in the name of efficiency. Fair labor practices involve setting achievable work standards based on accurate, humane evaluations and refraining from excessive automation that may displace workers without providing adequate retraining or support.

Environmental responsibility is another ethical domain. Industrial engineers are important in designing sustainable systems that minimize waste and pollution. When implementing lean practices or optimizing supply chains, they have a duty to assess and reduce environmental impacts. This can include selecting eco-friendly materials, designing energy-efficient processes, or reducing carbon footprints in logistics.

Additionally, industrial engineers must **maintain honesty and transparency** in their work, providing accurate assessments of projects' costs, risks, and benefits. Misrepresentation or cutting corners to reduce costs can lead to project failures, financial losses, or even harm to public safety. Professional bodies like the Institute of Industrial and Systems Engineers (IISE) offer ethical guidelines to help industrial engineers navigate these responsibilities.

In short, industrial engineers have a duty to prioritize people's welfare, promote environmental sustainability, and uphold transparency in every project. These ethical principles guide engineers to design systems that are not only efficient and profitable but also respectful of the individuals and communities they impact.

Emerging Trends and Future Directions in the Field

The field of industrial engineering is evolving rapidly, driven by technological advances, environmental challenges, and changing workforce dynamics. As organizations seek to operate more efficiently and sustainably, industrial engineers are adopting new methods, technologies, and mindsets to meet these demands. Several emerging trends and future directions are shaping the profession.

One significant trend is the integration of **Industry 4.0 technologies**, such as the Internet of Things (IoT), big data analytics, and cloud computing, into industrial engineering practices. IoT-enabled devices collect real-time data from equipment, machinery, and production lines, providing industrial engineers with insights that can optimize processes and predict maintenance needs. For instance, sensors can monitor equipment health, alerting engineers to potential breakdowns before they happen. Combined with big data analytics, these insights allow engineers to make data-driven decisions, reducing downtime and enhancing productivity.

Artificial intelligence and machine learning are also transforming industrial engineering. AI algorithms analyze data from production systems, supply chains, and customer interactions, identifying patterns and predicting future trends. Machine learning models can optimize production scheduling, quality control, and even inventory management by adapting to new information over time. These technologies enable predictive and prescriptive analytics, where industrial engineers can anticipate issues and recommend solutions rather than simply reacting to problems as they arise.

Sustainability is another prominent direction in industrial engineering. As businesses work to reduce their environmental impact, industrial engineers are adopting **green engineering principles** to create more sustainable processes. This includes designing for energy efficiency, reducing waste, and using renewable resources. Industrial engineers may also apply circular economy principles, designing systems that recycle materials back into the production process, minimizing the need for new resources. In logistics, for example, engineers are rethinking packaging, transportation routes, and warehouse layouts to reduce carbon footprints.

Human-centered design is gaining traction as workplaces become more automated. As robots and automated systems take on repetitive tasks, industrial engineers are focusing on creating environments that support worker engagement, creativity, and well-being. Ergonomic workspaces, flexible production lines, and intelligent machine interfaces allow humans and robots to collaborate effectively. This trend reflects a shift from purely task-oriented designs to those that prioritize employee experience and adaptability.

The rise of **remote work and virtual collaboration** is another future direction in industrial engineering, particularly in project management and global supply chains. Digital collaboration tools, virtual simulation models, and remote monitoring technology enable engineers to work on projects across different geographies, managing workflows and resources without being physically present. This shift not only offers flexibility but also allows organizations to draw on diverse expertise regardless of location.

As industrial engineers look to the future, **cybersecurity** has also become a critical consideration. With increasing reliance on digital networks and IoT, industrial systems are more vulnerable to cyber threats. Industrial engineers now have a role in protecting production systems from data breaches and ensuring secure data flows across supply chains. Implementing cybersecurity measures in IoT devices and production systems is essential to protect sensitive data and maintain operational integrity.

These trends indicate that the role of industrial engineers is becoming broader and more complex. As technology advances, industrial engineers will continue to adapt, using data, automation, and sustainability practices to create innovative solutions. These emerging trends are positioning industrial engineers not just as process optimizers but as forward-thinking designers of adaptable, resilient, and sustainable systems.

Industries and Applications: Where IE is Most Impactful

Industrial engineering has a profound impact across a wide range of industries, offering solutions that enhance efficiency, quality, and sustainability. By optimizing processes, designing effective systems, and improving resource use, industrial engineers make significant contributions to manufacturing, healthcare, logistics, retail, and service industries. Each of these sectors benefits uniquely from IE's methodologies and tools.

In **manufacturing**, industrial engineers focus on streamlining production processes to reduce costs, enhance quality, and minimize waste. They design assembly lines, optimize facility layouts, and implement lean manufacturing principles to improve productivity. For example, in the automotive industry, industrial engineers may analyze every step of the assembly process to ensure vehicles are produced with minimal waste and maximum quality. They use value stream mapping to identify inefficiencies, applying Six Sigma techniques to reduce defects and ensure that production meets high-quality standards. Additionally, industrial engineers are important in integrating automation into manufacturing, deploying robotics and IoT-enabled machines that increase precision and reduce manual labor requirements.

The **healthcare** sector has increasingly turned to industrial engineering to improve patient care, reduce wait times, and optimize resource use. In hospitals, industrial engineers work on scheduling systems to manage patient appointments, improve emergency room flow, and reduce bottlenecks in patient care. By applying lean techniques, they help hospitals reduce waste in processes like patient admissions and supply management. For instance, industrial engineers might streamline surgical scheduling to maximize operating room utilization and minimize patient wait times. They also contribute to facility layout design, ensuring that critical areas such as operating rooms, labs, and emergency departments are arranged for fast, efficient access.

In **logistics and supply chain management**, industrial engineers optimize the movement of goods from suppliers to customers, improving speed and reducing costs. They design warehouse layouts, optimize transportation routes, and use demand forecasting to manage inventory effectively. Retail giants, for example, rely on industrial engineers to structure their supply chains, ensuring that products are stocked in the right locations at the right times. Industrial engineers employ methods like the Economic Order Quantity (EOQ) model to balance ordering and holding costs, ensuring that inventory levels are maintained without overstocking. They also analyze data to plan efficient transportation routes, reducing delivery times and fuel consumption.

The **retail** industry uses industrial engineering to manage inventory, streamline operations, and improve customer service. Industrial engineers design stocking strategies, optimize store layouts, and manage workforce scheduling. For instance, in e-commerce, they might develop algorithms to forecast demand for specific items, ensuring popular products are available without excessive stock. They also work on improving warehouse operations, from automating order-picking processes to designing packaging solutions that maximize space and minimize shipping costs.

In **service industries** like banking, telecommunications, and hospitality, industrial engineers improve customer experience, reduce wait times, and optimize staff allocation. For example, in a call center, industrial engineers might analyze call patterns to forecast peak hours, ensuring adequate staffing to handle high call volumes. In hospitality, they may design workflows for hotel services, improving efficiency in areas such as check-in, housekeeping, and room service. By focusing on process efficiency and resource allocation, industrial engineers help service providers deliver a seamless experience that meets customer expectations.

Energy and utilities are also influenced by industrial engineering, particularly in optimizing resource use and improving system reliability. Industrial engineers in this field work on scheduling and maintenance planning to reduce downtime, as well as on energy-efficient design to reduce environmental impact. For example, in power plants, they may use predictive maintenance techniques to schedule repairs based on equipment condition, preventing costly breakdowns. Industrial engineers also contribute to the design of renewable energy facilities, such as wind and solar farms, by optimizing resource allocation, minimizing costs, and maximizing output.

In **transportation**, industrial engineers design systems to improve the efficiency of public transit, airports, and logistics networks. They work on reducing congestion, optimizing routes, and managing maintenance schedules for vehicles and infrastructure. For instance, in an airport, industrial engineers might streamline passenger flow from check-in to boarding, reducing bottlenecks and enhancing the traveler experience. They also develop scheduling systems for public transit, ensuring that vehicles are available during peak times and reducing wait times for passengers.

Overall, industrial engineers apply their skills across industries to design systems that are efficient, reliable, and adaptable. By reducing waste, improving quality, and optimizing resource allocation, they drive meaningful changes that benefit both organizations and their customers.

CHAPTER 2: PROCESS ANALYSIS AND DESIGN

Mapping and Documenting Processes

Mapping and documenting processes is a foundational task in industrial engineering. It involves visually laying out each step in a workflow to gain a clear understanding of how a process functions and where there might be inefficiencies or bottlenecks. This visual approach helps industrial engineers identify areas that need improvement, standardize operations, and establish a baseline for measuring future changes.

Process mapping starts with **identifying every step of a process**. Engineers begin by gathering information from people directly involved, such as operators, technicians, or managers, to understand the sequence of activities in detail. A successful map captures actions, decision points, resources used, and even potential errors. By including these elements, the map provides a comprehensive view of the process from start to finish. The goal is not only to document the steps but to see where value is added and where it's not, so that non-value-added steps (or waste) can be minimized.

One of the most common tools for this is a **flowchart**. Flowcharts use symbols to represent each step in a process, with arrows indicating the flow from one activity to the next. Each symbol—such as a rectangle for tasks, a diamond for decision points, or an oval for start and end points—provides clarity, making it easy to spot inefficiencies at a glance. For instance, if a decision point repeatedly leads to a delay or unnecessary loop in the process, it may indicate an opportunity to streamline or automate that decision. Flowcharts are simple but effective, and they are often the first tool industrial engineers turn to when they need a quick overview of a process.

A more detailed method is **Value Stream Mapping (VSM)**, often used in lean manufacturing. VSM goes beyond traditional flowcharts by including information flows, cycle times, and inventory levels. It shows not only the sequence of tasks but also where inventory is stored, how long each step takes, and whether each step adds value to the final product. For example, if a manufacturing process has several steps where materials sit in storage before the next step, VSM can highlight these delays. With this detail, industrial engineers can decide where to reduce wait times, optimize task sequences, or implement JIT (Just-in-Time) strategies to eliminate excess inventory.

Swimlane diagrams add another layer by dividing tasks across different departments or individuals. In processes involving multiple teams, like order fulfillment or customer service, swimlane diagrams help track which team handles each step, so it's clear when work changes hands. Each "lane" represents a person, team, or department, making it easier to identify where communication or coordination breaks down. For example, if a customer order passes from sales to fulfillment to shipping, each handoff point is highlighted. Engineers can then assess whether these transitions are causing delays or errors and explore options for smoother handoffs.

Another valuable method is **process flow analysis**. This technique documents the flow of materials, information, or actions across an entire process, including transportation and storage points. It is especially useful in environments like warehousing, where materials or products frequently change locations. By mapping out each stage, industrial engineers can determine if materials are moved more often than necessary. Reducing unnecessary movement not only saves time but also cuts costs associated with transportation and handling.

Throughout the mapping process, **data collection** is essential. Industrial engineers gather quantitative data, such as time spent on each task, frequency of errors, or inventory levels at each stage. This data enriches the process map, allowing for a more informed analysis. For instance, if the map shows that a particular machine has a high wait time before operators use it, data collection might reveal that scheduling or equipment maintenance is the cause. Without hard data, it's easy to make assumptions that may not hold up under closer scrutiny.

Once a process is mapped and documented, industrial engineers can proceed with a detailed **gap analysis**. This step involves comparing the current state with the desired or ideal state to identify performance gaps. If the goal is to reduce production time by 20 percent, for example, engineers examine the mapped process to find which steps can be shortened, automated, or removed entirely. By understanding these gaps, they can focus improvement efforts where they will have the greatest impact.

Mapping and documenting processes gives industrial engineers a structured, data-rich view of a system. It's the first step in identifying inefficiencies, standardizing operations, and finding opportunities for continuous improvement. With a well-documented map, industrial engineers can make informed changes that streamline workflows, cut costs, and increase productivity across various operations.

Identifying Bottlenecks and Inefficiencies

Identifying bottlenecks and inefficiencies is essential to improving any process. A bottleneck is a stage in a process where work accumulates or slows down, limiting the overall throughput. Inefficiencies, meanwhile, refer to unnecessary steps, wasted time, or resources that don't add value. By pinpointing these areas, industrial engineers can make targeted improvements that lead to smoother operations and higher productivity.

To identify bottlenecks, industrial engineers typically start by analyzing **process flow data**. They may gather quantitative data on time spent at each stage, waiting times, and work-in-progress (WIP) inventory levels. For example, if one workstation in a production line consistently has a longer cycle time than others, it's likely a bottleneck. Tracking WIP levels also provides clues; when inventory piles up at a particular stage, it indicates that downstream processes cannot keep up, causing a slowdown.

Cycle time analysis is another method for spotting bottlenecks. Cycle time is the time required to complete a specific task or step in a process. If one task has a notably higher cycle time than others, it may be creating a bottleneck. For instance, in a manufacturing process, if one machine takes twice as long to complete its job compared to other machines in the line, it will hold back the entire workflow. Engineers can address this by adding additional resources, adjusting task distributions, or reconfiguring workflows to match cycle times more evenly.

A bottleneck can also be identified by observing **idle times** in connected processes. When workers or machines have to wait for input from a previous stage, it indicates a blockage upstream. Engineers may monitor equipment utilization rates and downtime. Low utilization often means that a machine or worker is waiting due to delays elsewhere. In a call center, for example, if agents frequently wait for customer information from another department, that delay is an inefficiency that hinders productivity and adds to wait times.

Value Stream Mapping (VSM) is an effective tool for identifying both bottlenecks and inefficiencies. VSM provides a visual representation of the entire process flow, highlighting where resources are spent and how materials or information move through each stage. By mapping out each step, engineers can see where delays occur, where inventory accumulates, and which steps add value. If a particular step consistently slows down the process or contributes little to the final output, it may need to be streamlined or removed. VSM is often combined with data on cycle and lead times to provide a complete picture of where and why a bottleneck is occurring.

Root Cause Analysis (RCA) techniques are helpful when identifying why bottlenecks and inefficiencies exist. RCA involves examining all contributing factors, often using methods like the "5 Whys" or fishbone diagrams to explore different causes. For example, if a bottleneck occurs due to machine breakdowns, asking "Why?" multiple times may reveal that the root cause is infrequent maintenance or outdated equipment. By addressing the underlying cause rather than the symptom, engineers can eliminate the bottleneck more effectively.

In some cases, bottlenecks are **dynamic**; they move as changes are made to the process. When one bottleneck is resolved, another may emerge. For example, speeding up a specific task may shift the bottleneck to a downstream process that's unable to handle the increased volume. Industrial engineers need to monitor the entire system to anticipate how adjustments affect the overall flow. This is where **continuous monitoring and feedback loops** become crucial. By implementing systems to track key performance indicators (KPIs) in real time, engineers can identify new bottlenecks as they emerge and respond promptly.

In addition to bottlenecks, **inefficiencies** in workflows add waste without necessarily restricting throughput. Inefficiencies can occur in various forms, such as excess movement, redundant steps, or excessive handling of materials. Techniques like **motion study** and **time study** help identify where these inefficiencies lie. Motion studies observe workers and processes to eliminate unnecessary movements, while time studies measure how long each task takes, revealing areas that could be streamlined. In a warehouse, for example, if workers repeatedly walk long distances to retrieve items, changing the layout could reduce movement, saving time and energy.

Waste reduction techniques, especially those derived from lean manufacturing, also are important in identifying inefficiencies. The "seven wastes" in lean—overproduction, waiting, transport, extra processing, inventory, motion, and defects—provide a checklist for finding inefficiencies. For example, if a production process has significant waiting time between steps, it signals an inefficiency that could be addressed through better scheduling or parallel tasking.

Redesigning Processes for Optimal Flow

Redesigning processes for optimal flow focuses on restructuring workflows to reduce delays, remove bottlenecks, and eliminate inefficiencies. Optimal flow occurs when each task moves seamlessly from one step to the next, with minimal waiting time and no resource overloads. Achieving this requires a deep understanding of the current process, the challenges it faces, and how each part of the system interacts.

To start a redesign, industrial engineers analyze the **current workflow** to establish a baseline and understand existing pain points. This analysis often includes process mapping, time studies, and cycle time evaluations. By studying each step, they can determine where delays and interruptions occur. For instance, in a warehouse picking operation, analyzing order fulfillment can reveal steps where workers spend time searching for products or walking long distances. These findings then guide changes aimed at improving flow.

Once problem areas are identified, **process restructuring** can begin. One common technique is **workflow balancing**, which redistributes tasks among workers or machines to match their capacities more closely. In assembly lines, this often involves adjusting task loads to ensure each station completes its job within the same cycle time. For example, if one station takes longer than others, engineers may redistribute some of its tasks to nearby stations to prevent work from piling up. This balancing reduces idle time and keeps work moving continuously through the process.

In many processes, optimizing flow also involves **reducing handoffs and transitions**, which are points where delays commonly arise. Handoffs occur when work moves from one person or department to another, and they can create waiting times if the receiving party isn't prepared. To address this, industrial engineers might consolidate tasks, minimize touchpoints, or use parallel processing where possible. For instance, in a multi-step approval process, reducing the number of reviewers or combining approvals at one stage can save time and reduce interruptions, creating a faster, more direct workflow.

Automation is another approach to improving flow. Automated systems can handle repetitive tasks with greater speed and precision, reducing human error and freeing up employees for higher-value activities. In manufacturing, for example, robotic arms can handle assembly tasks consistently, minimizing delays associated with manual labor. Similarly, automated data processing in administrative workflows can replace manual data entry, speeding up tasks like order processing. However, industrial engineers must weigh the cost and complexity of automation against the expected improvements in flow.

Implementing **lean principles** also is significant in redesigning for optimal flow. Lean techniques aim to remove non-value-added activities and streamline processes. Tools like Value Stream Mapping (VSM) help industrial engineers identify waste and redesign workflows to focus only on essential tasks. By highlighting areas where time or resources are consumed without adding value, VSM guides changes that align tasks with customer needs. For example, in a production setting, engineers might eliminate unnecessary inspections or transport steps that don't directly contribute to product quality.

In some cases, **cellular manufacturing** offers an effective redesign solution. Cellular layouts group related tasks together in a cell, allowing products to move smoothly from one task to the next with minimal transport. This layout is especially beneficial for mixed-product lines or environments where flexibility is needed. Instead of moving parts across the entire factory, cells bring all necessary equipment and personnel close together, reducing movement and cutting down on production time. For example, a company producing customized products may create cells dedicated to each product line, making it easier to adjust to demand changes and avoid disrupting flow.

Bottleneck management is another key factor in designing for optimal flow. If certain tasks consistently delay production, redesign efforts focus on alleviating those bottlenecks. This may involve adding resources, such as extra machines or workers, to increase capacity at that stage, or reengineering tasks to be completed in less time. In a software development process, for instance, if code testing slows down production, increasing the number of testing tools or automating parts of the testing phase can accelerate the entire pipeline.

An essential aspect of process redesign is **continuous feedback and adjustment**. Optimal flow isn't a one-time fix; it's a dynamic target that changes as conditions evolve. Industrial engineers implement real-time monitoring systems and performance metrics to assess whether redesigned workflows achieve intended improvements. Metrics like cycle time, lead time, and throughput rate provide insights into how well the new process performs. For example, a manufacturing line might use digital sensors to track cycle times, allowing engineers to detect and respond quickly if any part of the workflow slows down unexpectedly.

Finally, **training and communication** are critical in implementing redesigned processes. Even the best-designed workflow won't perform optimally if workers aren't familiar with new procedures or equipment. Industrial engineers must ensure that team members understand changes and have the skills to execute tasks efficiently. Clear communication, hands-on training, and support materials help employees adjust to redesigned processes, ensuring a smooth transition and minimizing resistance.

Analyzing Process Cycle Times and Lead Times

Analyzing process cycle times and lead times is fundamental in industrial engineering, as these metrics provide insight into how efficiently a process operates. Cycle time and lead time, though often used interchangeably, measure different aspects of a process. **Cycle time** refers to the time taken to complete a single task or unit within a process, while **lead time** is the total time from the start of a process to its completion. By analyzing both, industrial engineers can identify bottlenecks, optimize workflows, and improve overall productivity.

To begin, cycle time analysis focuses on **individual tasks within the process**. Engineers break down each step and measure how long it takes to complete, starting from when a task begins to when it ends. This is particularly useful in assembly lines or repetitive processes where each step is dependent on the one before it. For example, in a packaging operation, cycle time analysis might reveal that placing items into boxes takes significantly longer than labeling. Knowing this, engineers can adjust the workflow, either by adding more workers at the packaging step or by exploring automation options to balance the cycle times across all steps.

Cycle time reduction is a common goal, as it directly impacts how many units can be completed within a given period. For instance, if the cycle time in a production line is reduced from 10 minutes to 8 minutes, this change increases the output by 20% over the same period. However, reducing cycle time requires careful consideration of quality and safety; cutting time by compromising quality is counterproductive and can lead to increased waste or rework.

In addition to analyzing individual steps, **lead time analysis** provides a broader perspective. Lead time includes all aspects of the process, from the initial request or order to the final delivery of the product or service. This metric accounts for both active work time and any delays, such as waiting for materials, approvals, or transportation. For example, in a supply chain, lead time measures the full duration from when a customer places an order to when they receive the product, covering procurement, production, and shipping times.

Lead time analysis is crucial for **understanding customer impact**, as it reflects how long customers must wait for their orders. If lead times are consistently long or unpredictable, customers may become dissatisfied, and the company may struggle with inventory management or lost sales. By examining lead times, industrial engineers can identify specific stages where delays occur and determine if resources are being underutilized or if external factors, like supplier delays, are impacting the process.

To gather data for cycle and lead time analysis, industrial engineers may use tools like **time studies**, in which each step is observed and recorded to calculate average times accurately. Digital monitoring tools, such as sensors and automated trackers, are also valuable, especially in production environments where real-time data can reveal fluctuations in cycle times. This data is essential for setting accurate baselines and for monitoring improvements over time.

One of the main objectives of analyzing cycle and lead times is to identify **bottlenecks**. A bottleneck is a stage in the process where work accumulates, slowing down the overall flow. If a particular step consistently has a longer cycle time than others, it can create a backlog that increases the lead time for the entire process. Addressing bottlenecks often requires reallocating resources, balancing workloads, or even redesigning the workflow. For instance, if an approval process in an administrative workflow delays completion, industrial engineers might adjust responsibilities or automate parts of the approval to accelerate this step.

Calculating takt time can also help in synchronizing cycle times with demand. Takt time is the rate at which products need to be completed to meet customer demand. By comparing takt time with actual cycle times, industrial engineers can assess whether the current production rate aligns with customer expectations. If cycle times exceed takt time, it indicates that the process is too slow, and adjustments are necessary to meet demand.

Finally, tracking **improvements over time** is essential to gauge the effectiveness of any changes made. For instance, if lead time reductions are a target, monitoring before-and-after lead times reveals whether interventions like supplier adjustments, process redesigns, or scheduling changes were effective. Similarly, a reduction in cycle time following the addition of automation confirms that changes have successfully increased throughput.

CHAPTER 3: WORK MEASUREMENT AND TIME STUDIES

Quantifying Work for Efficiency

Quantifying work is essential for improving efficiency in any operation. In industrial engineering, quantifying work involves measuring the time and effort required to complete tasks, setting standards, and identifying areas for improvement. When work is quantified accurately, engineers can determine how best to allocate resources, set realistic targets, and design processes that maximize productivity while minimizing waste.

One of the most common methods to quantify work is through **time studies**. In a time study, an industrial engineer observes a task and measures how long it takes to complete under normal working conditions. This study involves dividing a task into smaller, measurable steps and timing each one individually. For example, if a worker assembles parts on a production line, the time study might break down each step—from picking up materials to securing parts together—to find the total time needed for the entire assembly. Time studies not only provide an accurate measure of the time required but also highlight any unnecessary movements or delays that may be adding time without adding value.

Work sampling is another tool used to quantify work. Unlike time studies, where tasks are continuously observed, work sampling involves observing tasks at random intervals over a period. This method provides data on how much time workers spend on various activities, such as productive work, waiting, or idle time. For instance, in a warehouse setting, work sampling can reveal how much time is spent on picking items, transporting them, or waiting for new instructions. By understanding where time is spent, engineers can make informed decisions to improve workflow and reduce downtime.

Establishing **standard times** is a key outcome of work measurement. A standard time is the expected time for a qualified worker to complete a task under normal conditions. To calculate standard time, industrial engineers often take the observed time from a time study and adjust it for performance ratings and allowances (like fatigue or unavoidable delays). For example, if an assembly task is observed to take 10 minutes, and the performance rating is 90%, the adjusted time becomes 11.1 minutes. This standard time helps set realistic productivity targets and guides staffing, scheduling, and resource allocation.

Predetermined Motion Time Systems (PMTS) provide another way to quantify work by assigning time values to specific motions or actions. PMTS breaks down tasks into basic motions, such as reaching, grasping, or positioning, and assigns a time value to each based on the conditions. By summing these values, engineers can estimate how long a task should take without direct observation. For example, a

PMTS like MTM (Methods-Time Measurement) assigns time units to each motion, so an engineer can calculate the time required for an assembly task involving several precise hand movements. PMTS is especially useful for setting standards in repetitive tasks where consistent time estimates are needed.

Once work is quantified, **efficiency analysis** begins. Industrial engineers compare actual performance against the standard time to assess productivity. If a task consistently takes longer than its standard time, it may indicate a need for changes, such as ergonomic adjustments, better tools, or additional training. In contrast, if a task is completed well under the standard time, it may mean the standard needs revision or the process is working exceptionally well. This ongoing comparison allows organizations to stay agile and continually optimize processes based on real data.

Quantifying work also supports **resource planning and allocation**. When engineers know how long tasks should take, they can determine how many workers, machines, or workstations are needed to meet production goals. In a call center, for example, knowing the average time to handle a call helps managers estimate the required number of agents during peak hours, ensuring service levels are maintained without overstaffing. This precise allocation of resources prevents wasted time and maximizes productivity.

Incentive systems are often based on quantified work measurements. Organizations may set up performance-based incentives that reward workers who exceed standard times while maintaining quality. For example, if a worker consistently assembles units faster than the standard time, they might receive a bonus. These incentives encourage efficiency, but they must be managed carefully to avoid burnout or compromised quality.

Overall, quantifying work is about creating a structured, data-driven approach to understand and improve productivity. By measuring work accurately and consistently, industrial engineers ensure processes are both efficient and adaptable, helping organizations meet their goals in a sustainable way.

Time Study Techniques

Time study techniques are essential in work measurement, allowing industrial engineers to accurately measure how long tasks take under normal working conditions. By analyzing these measurements, engineers can establish realistic standards for productivity, assess performance, and identify areas where processes can be improved.

A **direct time study** is one of the most commonly used methods. In this approach, the engineer observes an operator performing a task and uses a stopwatch to record the time taken for each element of the task. The study typically involves breaking down the task into discrete elements, such as "pick up part," "align with fixture," or "secure with screw." Each element is timed separately, and the process is repeated multiple times to account for variability and ensure accuracy. For instance, if assembling a product requires five distinct steps, each step is timed separately across several cycles to obtain an average time for each. This breakdown allows engineers to pinpoint specific steps that may be consuming excessive time, offering clear insights for improvement.

A key aspect of direct time studies is **performance rating**. Performance rating involves evaluating the operator's speed relative to a defined standard, often set at a pace considered "normal" or average. If an operator works faster or slower than this standard pace, the recorded time is adjusted to reflect the expected time for an average worker. For example, if an operator completes a task faster than the normal rate, the measured time might be increased proportionally to reflect a realistic time for an average pace. This adjustment ensures that the resulting time standard is fair and achievable by a typical worker.

In addition to direct observation, some engineers use **video-based time studies**. This method involves recording the task on video and then analyzing it in detail afterward. By playing back the recording, engineers can measure time intervals more precisely and analyze elements that may be difficult to capture in real time. Video studies are particularly useful for complex or fast-paced tasks where direct observation may not capture every detail accurately. They also enable team members to review the footage together, allowing for collaborative analysis and insight.

Work sampling is another technique used to estimate time spent on different activities without continuous observation. Instead of timing every task directly, the engineer takes random "snapshots" of work throughout the day to record what activity is being performed at each point. For example, in a factory, work sampling might involve checking in on workers every 10 minutes to record their activities, providing an estimate of time spent on productive versus non-productive tasks. This method is especially valuable in environments where direct time studies are impractical, like in administrative work or when studying multiple workers across different locations.

Allowances for fatigue, personal needs, and unavoidable delays are often added to time studies to make standards more realistic. These allowances account for the fact that workers may need breaks, experience minor delays, or have some personal time during the workday. If a task cycle time is determined to be five minutes and the fatigue allowance is set at 5%, the final standard time would be adjusted to 5.25 minutes. Allowances ensure that time standards remain fair and attainable, factoring in the realities of the work environment.

Each of these techniques serves a specific purpose, providing detailed and precise measurements that allow for accurate work standards. Time studies provide a foundation for assessing productivity, designing efficient workflows, and setting clear expectations for performance.

Standard Data and Predetermined Motion Time Systems

Standard data and predetermined motion time systems (PMTS) are critical elements in work measurement, providing standardized time values for common tasks and motions. By using these established data, industrial engineers can set accurate time standards without needing to conduct time studies for every task, making the process of work measurement more efficient and consistent.

Standard data refers to pre-established time values for specific tasks or elements within tasks, based on past time studies. For instance, if an organization has conducted multiple time studies on similar tasks, it may develop standard data for common actions like "picking up an object" or "walking 10 feet." These data points become benchmarks that engineers can apply to new tasks without needing to re-measure each action. For example, in a production facility, if data shows that tightening a screw takes an average of five seconds, this value can be reused in similar tasks across different products. Standard data saves time and provides a reliable basis for setting time standards, especially in environments with high task repetition.

Predetermined Motion Time Systems (PMTS), such as MTM (Methods-Time Measurement) or MOST (Maynard Operation Sequence Technique), take standard data further by breaking down tasks into basic motions, each assigned a precise time value. These systems assign time to fundamental actions, like reaching, grasping, or positioning, based on extensive studies of human movement. For example, MTM may define a "reach" as taking 0.1 seconds under certain conditions, and a "grasp" as taking 0.08 seconds. By summing these values for each motion involved in a task, engineers can calculate the total time required without direct observation. PMTS are particularly useful in assembly line settings, where tasks involve repetitive motions and need consistent time standards across operations.

The advantage of PMTS is that it eliminates variability and observer bias, providing a **precise and objective method** to set time standards. Since each motion is pre-assigned a value, the resulting time estimate is consistent across similar tasks and operators, regardless of individual performance variations. In a scenario where an operator needs to pick up, move, and assemble a part, PMTS allows engineers to calculate the total time by adding the times for each motion—reach, grasp, move, and position—yielding a reliable time standard that doesn't rely on real-time observation.

Using PMTS also enables **rapid analysis of complex tasks**. For example, in a manufacturing environment where a product involves multiple assembly steps, engineers can break down each step into basic motions and sum the values. This analysis makes it possible to estimate the time required for new tasks, even before production starts. PMTS helps speed up the process of setting standards for new products, enabling faster ramp-up times and reducing the need for extensive time studies on-site.

While PMTS provides accuracy and consistency, **allowances** are still added to account for factors like fatigue, personal time, and delays. Since PMTS calculates ideal times based solely on motion data, it doesn't account for real-world interruptions. Allowances, often ranging from 5% to 15%, adjust the standard times to be more realistic, ensuring that expectations are achievable in actual working conditions.

In environments with varied tasks, **combining standard data and PMTS** can be effective. Engineers might use standard data for tasks with predictable outcomes and PMTS for more detailed analysis of high-frequency or repetitive tasks. Together, these tools create a comprehensive framework for establishing reliable time standards that support productivity goals and streamline operations without excessive reliance on new time studies.

CHAPTER 4: FACILITY LAYOUT AND MATERIAL HANDLING

Principles of Effective Facility Layout

Designing an effective facility layout is essential for smooth operations, minimizing material handling, and maximizing productivity. In industrial engineering, the layout determines how people, machines, and materials are arranged to optimize flow and support efficient production. Good layout design considers process flow, ease of movement, and safety, creating an environment where work is done efficiently with minimal waste.

A **principle of effective layout** is proximity, which places related tasks or equipment close to each other to minimize movement and reduce transportation time. This principle is especially critical in production environments where materials move through multiple steps. For example, in a manufacturing plant, arranging workstations so that raw materials flow logically from one operation to the next—cutting, welding, assembly—reduces unnecessary movement, saving time and lowering labor costs. Each step's location is carefully chosen to ensure a smooth transition, minimizing transport needs.

Flow efficiency is another core principle. Industrial engineers design layouts to ensure materials, products, and people can move seamlessly without bottlenecks. Efficient flow means there are few interruptions or backtracking in a production process. For instance, a U-shaped or straight-line layout often supports efficient flow, allowing workers and materials to move linearly without crossing paths. An effective layout achieves a balance between keeping everything close together and ensuring that each task has adequate space. Overcrowding slows down operations and increases the risk of accidents, while excessive spacing wastes time and energy.

Flexibility is crucial in layout design, particularly in environments where production needs can change rapidly. Flexible layouts allow for quick reconfigurations to accommodate new products, varying production volumes, or process changes. For example, modular workstations or mobile equipment can be rearranged without major disruptions, enabling the facility to adapt to demand changes. This flexibility is particularly beneficial in industries with frequently changing product lines, such as electronics, where each model may require a different assembly setup.

Safety is integral to an effective facility layout. Engineers must design layouts to minimize hazards, providing clear paths for workers, equipment, and materials. **Safety zones** around high-risk areas, like heavy machinery or welding stations, ensure that only authorized personnel can access these areas, reducing accident risk. Pathways for forklifts or carts should be separated from pedestrian areas to avoid collisions, with clear markings and signage guiding traffic flow. Additionally, exit routes and emergency equipment like fire extinguishers must be strategically located and easily accessible.

Effective layouts also incorporate **ergonomics** to enhance worker comfort and reduce strain. Workstations are designed to support natural movements, minimizing bending, reaching, or twisting. For example, in an assembly line, placing tools and materials at waist height allows workers to perform tasks with minimal effort. Ergonomic layouts reduce fatigue and prevent repetitive strain injuries, directly impacting productivity and quality. In environments where workers perform similar tasks throughout the day, ergonomic design keeps them comfortable and efficient.

Minimizing material handling is a key consideration in facility layout. Material handling is often a non-value-added activity, meaning it doesn't directly contribute to the final product. Reducing the distance materials travel and the number of handling steps improves efficiency and reduces costs. Engineers analyze the flow of materials through the layout to locate heavy or frequently used items near their point of use, eliminating excess transport. For instance, a facility might place storage areas for raw materials adjacent to processing areas, minimizing the need to move materials over long distances.

Storage optimization is another principle of effective facility layout. Engineers carefully plan where to place inventory, tools, and raw materials to prevent overcrowding and maintain an organized environment. Efficient storage design ensures that items are easily accessible when needed but do not interfere with production flow. For example, in a warehouse, high-demand items are often stored at ground level or close to shipping areas, minimizing retrieval time.

Lastly, **visual management** principles enhance layout effectiveness by providing clear, visible information about processes, equipment, and flow paths. Signs, labels, and color-coded areas make it easy for workers to navigate the facility, find tools, and follow safety protocols. Visual cues streamline operations by reducing confusion, preventing errors, and reinforcing the intended flow of work. For instance, color-coded pathways can indicate safe walking areas, material handling zones, or restricted areas, ensuring that all employees understand the layout intuitively.

Types of Layouts: Product, Process, and Cellular

Choosing the right layout type is essential in facility design as it directly affects workflow, material movement, and efficiency. Three primary layout types—product, process, and cellular—each suit different production requirements, offering unique advantages based on the needs of the operation.

A **product layout**, also known as a line layout, arranges workstations and equipment in a linear sequence to match the production process for a single product or a limited range of similar products. This layout type is ideal for high-volume, low-variety manufacturing, such as automotive assembly lines. In a product layout, each workstation completes a specific task, and the product moves sequentially from one station to the next without backtracking. For example, in a car assembly plant, the car frame starts at one end and moves down the line where each station handles specific components—engine installation, body assembly, painting, and so forth. The main advantage of this layout is its efficiency in producing large volumes consistently. With each task standardized and sequentially organized, the workflow is smooth, minimizing delays between stations.

However, product layouts lack flexibility, as they are optimized for specific tasks in a fixed order. Any variation in product design or production volume changes can disrupt the flow and require significant reconfiguration. Downtime in one station can also halt the entire line, so maintenance and smooth handoffs between stations are critical to keep the flow uninterrupted. These layouts are best suited for industries where consistency and high output are essential.

Process layouts, or functional layouts, group similar processes or machines in designated areas. This layout is common in settings where a variety of products or custom work is required, like in machine shops or hospitals. In a process layout, each area focuses on a specific type of operation (e.g., drilling, painting, or assembly), and products move between stations based on their unique processing needs. For instance, in a machine shop, lathes, milling machines, and drill presses are grouped separately, and each product follows a path that matches its production requirements.

Process layouts are highly flexible, allowing facilities to handle diverse products and adapt quickly to changes in customer demand. Because products don't follow a single path, they can move as needed between different stations, which is particularly useful for low-volume, high-variety production. However, the flexibility of process layouts can lead to more complex material handling requirements, with products potentially moving back and forth across the facility. This layout type also increases waiting times and setup changes between tasks, making it less efficient for high-volume production compared to a product layout.

Cellular layouts combine elements of both product and process layouts, grouping workstations and equipment into cells that focus on producing a family of similar products. Each cell is arranged like a mini production line, with all necessary equipment positioned to complete a specific set of related tasks. In a furniture factory, for instance, a cell might include stations for cutting, sanding, and assembly specifically for a line of chairs, allowing each product family to be produced in a continuous, efficient flow.

Cellular layouts increase flexibility while retaining some efficiency benefits of product layouts. Each cell operates independently, allowing for more adaptable production schedules and reducing movement between different process areas. Cellular manufacturing also shortens lead times, as work can flow through each cell with minimal interruptions. Additionally, workers in cellular layouts often have cross-training to handle multiple tasks, which improves productivity and enhances adaptability. However, cellular layouts require careful planning and may need additional investment to set up cells effectively for each product family.

Each layout type has specific strengths and limitations, making the choice dependent on production goals, product diversity, and facility size. Product layouts prioritize efficiency for high-volume, single-product lines, while process layouts provide flexibility for varied, low-volume production. Cellular layouts strike a balance, enabling efficient, family-oriented production without sacrificing adaptability.

Material Handling Techniques

Efficient material handling is essential in facility design, as it directly impacts productivity, cost, and safety. Material handling involves the movement, storage, control, and protection of materials throughout the production process, from raw materials entering the facility to finished goods leaving it. Effective material handling techniques streamline these movements, reducing delays and minimizing costs associated with unnecessary handling.

Manual handling is the most basic material handling method, where workers move, lift, or carry materials by hand. While it's simple and low-cost, manual handling has limitations, especially for heavy or bulky materials. Relying on manual handling for extensive movements can lead to slower operations and increase the risk of injury. To mitigate these risks, industrial engineers design facilities to limit manual handling, introducing tools like carts or conveyors to help workers move materials with minimal effort.

Conveyor systems are commonly used to automate material movement along fixed paths. Conveyors are especially valuable in production lines where materials need to move between workstations consistently. For example, in a bottling plant, conveyors move bottles from the filling station to capping and labeling areas. Conveyors improve flow and eliminate the need for manual transport between stations, speeding up operations and reducing labor costs. However, conveyors work best in environments with predictable, linear workflows, as they're limited by fixed paths and require a structured setup to operate efficiently.

For facilities needing more flexibility in movement, **forklifts and pallet jacks** provide versatile options. Forklifts can transport heavy loads over short to medium distances and are widely used in warehouses, loading docks, and manufacturing facilities. Unlike conveyors, forklifts are mobile, allowing them to transport materials across variable routes as needed. For smaller loads or shorter distances, pallet jacks provide a simpler, cost-effective solution for moving pallets. However, forklifts and pallet jacks require skilled operators and are best suited for facilities with open, navigable spaces where they can move freely.

Automated Guided Vehicles (AGVs) offer a more advanced, automated approach to material handling. AGVs are robotic vehicles programmed to follow specific paths within a facility, transporting materials from one location to another without human intervention. They can move along preset paths marked by magnetic strips, sensors, or lasers, making them ideal for repetitive tasks or environments that benefit from automated, consistent movement. For instance, in an automotive parts factory, AGVs might transport parts between assembly lines and storage areas. While AGVs reduce labor costs and increase safety, they involve a higher upfront investment and are typically used in larger facilities with high material handling needs.

In environments where **storage and retrieval are frequent**, automated storage and retrieval systems (AS/RS) enhance material handling efficiency. AS/RS uses cranes, shuttles, or other automated machinery to retrieve materials from storage and deliver them to workstations. These systems are particularly effective in high-density storage areas, such as warehouses, where space is limited. By reducing the need for manual retrieval, AS/RS speeds up operations and maximizes storage efficiency. However, AS/RS also requires careful planning and a structured layout to ensure seamless integration into the facility's workflow.

Material flow planning is integral to these techniques, as it ensures each method aligns with the facility layout and operational needs. Engineers analyze the movement paths of materials, aiming to minimize unnecessary handling and movement. For example, in a process layout where materials flow between various functional areas, a combination of forklifts and conveyors might support flexibility and efficiency. In a product layout, fixed conveyors often suffice, as materials move in a predictable, linear sequence.

Integrating Automation in Material Flow

Integrating automation in material flow transforms how materials move through a facility, reducing manual labor, improving efficiency, and minimizing errors. When implemented effectively, automation allows materials to flow seamlessly between workstations, storage, and shipping, enhancing the overall productivity of a facility. Industrial engineers design facility layouts to incorporate automation based on the production needs, process types, and facility size.

Conveyor systems are a foundational component in automated material flow, especially in environments with high volumes and predictable movement patterns. Conveyors move materials consistently along fixed paths, ensuring that materials travel between workstations with minimal delays. In a food processing plant, for example, conveyor systems move products from processing to packaging, maintaining a continuous, reliable flow without manual intervention. These systems also support automated sorting, enabling items to be directed to specific destinations based on criteria such as weight, shape, or barcode data. Automated sorting reduces handling time and ensures that materials reach the correct location, reducing the potential for misplacement or delays.

Automated Guided Vehicles (AGVs) offer flexibility in automated material handling. Unlike conveyors, AGVs can move freely around the facility, following programmed paths marked by magnetic strips, sensors, or laser guidance systems. AGVs are particularly useful in facilities with variable workflows or complex layouts where materials need to reach multiple points. For instance, in a large automotive plant, AGVs might transport parts from the warehouse to different assembly stations based on production schedules. Engineers program AGVs to operate on specific routes or respond dynamically to traffic within the facility, ensuring they avoid obstacles or congestion. By integrating AGVs, facilities can handle variable material flows without needing fixed infrastructure, making AGVs ideal for adaptable layouts and mixed-use spaces.

Robotic arms and other robotic equipment are important in automating material flow within workstations. In tasks like assembly, sorting, or packing, robotic arms perform repetitive actions with precision, significantly reducing cycle times. For example, in a pharmaceutical packaging facility, robotic arms may sort and place products into boxes, eliminating manual labor and increasing speed. These robots are programmable and can handle complex motions, such as rotating, lifting, and placing items. This integration with automated material flow ensures a smooth transition from one task to the next, minimizing interruptions and maintaining a steady throughput.

Automated Storage and Retrieval Systems (AS/RS) are highly effective in warehouses and storage-intensive facilities. AS/RS uses cranes, shuttles, or robotic arms to store and retrieve materials automatically, reducing the need for manual picking. In e-commerce warehouses, for instance, AS/RS locates and retrieves items based on real-time demand, optimizing order fulfillment speed and reducing labor costs. Engineers often position AS/RS systems near production or shipping areas to minimize travel distances and ensure quick access to materials. With AS/RS, facilities can maximize storage density and improve material flow without the delays or errors associated with manual retrieval.

Integration with software systems like Warehouse Management Systems (WMS) and Enterprise Resource Planning (ERP) further enhances automated material flow. These systems provide real-time data on material locations, inventory levels, and demand forecasts, allowing automation to respond dynamically to changing requirements. In a distribution center, for instance, WMS software coordinates AGVs, AS/RS, and conveyors based on order priorities and inventory levels. The WMS can reroute AGVs or adjust conveyor speeds based on real-time demand, optimizing material flow and ensuring timely order fulfillment. By integrating automation with software systems, facilities achieve a higher degree of control and adaptability, enabling automated processes to respond intelligently to workflow variations.

Automated inspection and quality control systems also are important in material flow, especially in high-precision industries. Vision systems and sensors inspect materials as they move through conveyors or AGVs, identifying defects or discrepancies in real time. In electronics manufacturing, automated vision systems check components for accuracy as they pass through assembly lines. If defects are detected, automated sorting mechanisms remove the faulty items from the line, ensuring only quality materials continue through the process. This approach reduces manual inspection needs, improves consistency, and maintains material flow by eliminating defective items before they enter further stages of production.

Integrating automation in material flow requires a comprehensive approach, where facility layout is designed to support automated paths, stations, and storage. By coordinating conveyors, AGVs, robotic arms, AS/RS, and software systems, industrial engineers create a streamlined, automated flow that enhances facility performance and reduces the need for manual handling across operations.

CHAPTER 5: LEAN MANUFACTURING PRINCIPLES

Understanding Lean and its Core Concepts

Lean manufacturing is a production philosophy centered on maximizing value while minimizing waste. Developed from the Toyota Production System, lean focuses on identifying and eliminating any process steps that do not add value from the customer's perspective. Lean principles guide companies to streamline operations, cut costs, and improve quality by removing inefficiencies, which ultimately leads to faster production times and a more responsive operation.

A core concept in lean is the **elimination of waste**, known in Japanese as *muda*. Lean identifies seven primary types of waste: overproduction, waiting, unnecessary transport, overprocessing, excess inventory, unnecessary motion, and defects. Each type of waste detracts from value and adds cost without benefiting the customer. For example, overproduction—producing more items than needed—leads to excess inventory that takes up storage, increases handling, and can lead to obsolete products. Lean manufacturing continually seeks to remove these wasteful elements, focusing resources only on steps that create value for the customer.

Value stream mapping is a key tool in lean, used to visualize every step in a process and highlight where waste occurs. Engineers map out the flow of materials, information, and people within a process, identifying non-value-added activities. For instance, if mapping a process reveals multiple delays between steps, engineers can investigate whether steps could be combined or if scheduling improvements could reduce waiting times. By creating a value stream map, lean practitioners gain a clear view of the current process, making it easier to pinpoint areas where time, materials, or labor are being wasted.

Just-in-Time (JIT) production is another core concept in lean, aimed at reducing inventory levels by producing only what is needed, when it's needed. JIT minimizes storage costs and reduces the risk of overproduction, as items are manufactured or assembled in response to customer demand. In a JIT environment, suppliers deliver parts and materials just as they are required on the production line, creating a continuous flow. This requires strong coordination with suppliers and tight control over scheduling. For example, an auto manufacturer using JIT would receive parts like engines or seats precisely when assembly calls for them, rather than holding large quantities in stock.

Closely tied to JIT is the concept of **pull systems**. In traditional manufacturing, production is often a "push" system where items are made based on forecasts, which can lead to surplus and excess inventory. In lean, however, a pull system dictates that each step in production starts only when the next step is ready to receive it, reducing unnecessary buildup of materials. For instance, in a kanban system—a pull technique often used in lean—cards or signals indicate when more parts are needed, ensuring that resources are replenished based on real demand rather than anticipated demand.

Continuous improvement, or *kaizen*, is foundational in lean, promoting small, incremental changes to improve processes continuously. Kaizen encourages employees at all levels to look for ways to make their work easier, faster, or more effective. These improvements are not about large, one-time overhauls but focus on consistent, manageable changes. For example, in a lean production setting, operators might find a way to reposition tools within reach, shaving seconds off each task. While it may seem minor, these changes add up significantly over time, enhancing efficiency and supporting a culture of proactive problem-solving.

Standardized work ensures consistency and quality in lean operations. By defining the most efficient way to perform each task and documenting it as a standard, organizations reduce variation, minimize errors, and make it easier to identify opportunities for improvement. For instance, in an electronics assembly line, standardized work might dictate a precise sequence for installing components, with each step timed and optimized. Standardizing tasks not only improves quality but also allows new employees to train faster and adapt to their roles with less supervision.

Another critical concept in lean is **5S**, a workplace organization method designed to improve cleanliness and order, making it easier to spot issues and keep processes running smoothly. The 5S system—Sort, Set in order, Shine, Standardize, and Sustain—ensures that tools and materials are organized for efficiency. For example, in a workshop, 5S might mean clearing out unused tools, setting up labeled storage areas, and implementing regular cleaning routines. 5S promotes a tidy, organized workspace where everything has a place, reducing time wasted searching for tools and improving safety.

Together, these lean concepts create a structured, data-driven approach to improving production processes. By focusing on eliminating waste, continuous improvement, and just-in-time production, lean manufacturing fosters a responsive, efficient environment that aligns every activity with value creation for the customer.

Identifying and Reducing Waste (Muda)

In lean manufacturing, **muda** is the term used for waste—any activity that consumes resources without adding value from the customer's perspective. Identifying and reducing these wastes is essential for creating a streamlined, efficient operation. Lean manufacturing identifies seven types of waste that can hinder productivity, increase costs, and reduce product quality.

The first type of waste is **overproduction**, which occurs when products are made before they are needed. This leads to excess inventory, taking up storage space and requiring additional handling. For example, producing 1,000 units when only 700 are needed results in surplus inventory that may become obsolete or require costly storage. Reducing overproduction aligns manufacturing with actual demand, preventing the cost and risks of excess stock.

Waiting is another common form of waste, occurring when resources sit idle between tasks. This can happen due to delays in receiving materials, equipment downtime, or bottlenecks in the workflow. For instance, if a worker waits for parts to arrive from a previous station, the downtime is non-value-added. Reducing waiting time requires identifying and smoothing out bottlenecks, improving scheduling, and ensuring resources are available precisely when needed.

Transportation waste involves unnecessary movement of materials within a facility. Moving materials multiple times adds no value and increases the risk of damage. For example, moving a part across a warehouse to another area for processing increases handling time without enhancing the final product. Simplifying layouts and locating workstations closer together can reduce this waste, leading to faster material flow.

Overprocessing refers to using more resources or performing more work than necessary to achieve the desired result. This might include adding extra finishing steps or inspections that don't add significant value to the customer. For example, in a machining operation, unnecessary polishing or redundant inspections add time and cost. To address overprocessing, engineers should ensure that each step is essential and streamlined to meet customer requirements without excess.

Excess inventory is another waste, often linked to overproduction. Storing large amounts of raw materials or finished goods ties up capital and requires additional storage space. It also increases the risk of product obsolescence. Implementing just-in-time (JIT) practices and carefully managing supply chain logistics can help keep inventory levels aligned with actual demand, reducing unnecessary holding costs.

Motion waste includes unnecessary movements made by workers, such as reaching, bending, or walking long distances to retrieve tools or materials. Motion waste not only consumes time but can also lead to fatigue and injury over time. For example, if a worker has to walk back and forth to retrieve tools for a task, rearranging the workstation to keep tools within reach can reduce motion waste, improving both productivity and ergonomics.

Finally, **defects** represent one of the most costly forms of waste, as they require rework or result in scrapped products. Defects occur due to errors in production, faulty materials, or inadequate quality control. For example, in an assembly line, if parts are frequently misaligned, it results in additional inspection, rework, or disposal costs. To minimize defects, lean practices emphasize root cause analysis and continuous quality improvement.

Techniques for Just-in-Time (JIT) Production

Just-in-Time (JIT) production is a lean manufacturing strategy that focuses on producing only what is needed, when it's needed, and in the quantities required. JIT aims to minimize inventory, reduce waste, and improve efficiency by aligning production schedules closely with actual demand. Several techniques support the effective implementation of JIT in manufacturing environments.

Kanban is one of the primary tools used in JIT systems. It is a visual scheduling method that uses cards, bins, or digital signals to indicate when a specific part or material is needed. Each Kanban card corresponds to a batch of parts or materials, moving through the production line as needed. For instance, if a workstation completes a batch, it sends a Kanban signal to the previous station to supply more materials. This pull-based system prevents overproduction by producing items only in response to demand, ensuring that each stage only works when needed.

Setup time reduction is crucial in JIT production, as it allows for smaller batch sizes and more frequent changeovers without excessive downtime. Techniques like SMED (Single-Minute Exchange of Dies) help reduce setup times by organizing tools, using quick-change fixtures, and streamlining procedures. For example, in an automotive assembly line, SMED techniques enable workers to switch from one car model to another within minutes, maintaining flow without building up excess inventory. Setup time reduction provides flexibility, allowing manufacturers to respond quickly to changes in demand without accumulating surplus.

Production leveling, or *Heijunka*, is another JIT technique focused on distributing production evenly over time to meet demand without overloading certain production stages. In practice, Heijunka balances the workload by producing a mix of products or varying quantities based on real-time demand. For instance, instead of producing all units of one product before switching to another, production is scheduled in smaller, mixed batches. This steady production flow reduces lead times, prevents bottlenecks, and helps maintain a consistent pace that aligns with customer demand.

Supplier partnerships are essential in JIT systems to ensure that materials arrive on time and in the right quantities. JIT relies heavily on a dependable supply chain, as there is minimal inventory buffer. Many companies form close, collaborative relationships with suppliers, setting up frequent deliveries of smaller quantities to align with production needs. For instance, a car manufacturer might schedule daily or even hourly deliveries of components, coordinating with suppliers to avoid holding excess stock. Reliable suppliers are critical to JIT success, as any disruption can immediately affect production flow.

Automation and real-time monitoring support JIT by providing precise control over production schedules and inventory. Automated systems track inventory levels, production progress, and demand changes, allowing for adjustments in real time. For example, a factory using IoT-enabled sensors and ERP (Enterprise Resource Planning) systems can monitor raw material levels and trigger orders or production changes automatically. This technology reduces the risk of stockouts or overproduction by keeping production in sync with actual demand.

Employee training and involvement are also integral to JIT, as the approach requires a proactive, responsive workforce. Workers are trained to identify issues early, maintain equipment, and contribute ideas for process improvement. For instance, in a JIT environment, operators may suggest adjustments to workstations that minimize setup times or propose improvements to address quality issues quickly. Employee involvement enhances the adaptability of the system and ensures that each part of the production line can respond efficiently to changes in demand.

JIT techniques work together to create a responsive, efficient production system that minimizes inventory and aligns closely with actual demand. By implementing Kanban, reducing setup times, partnering with reliable suppliers, and leveraging automation, companies can achieve a streamlined, demand-driven operation.

Value Stream Mapping in Lean Systems

Value Stream Mapping is a popular framework in lean manufacturing used to visualize, analyze, and improve the flow of materials and information within a process. VSM helps identify inefficiencies, bottlenecks, and areas of waste by capturing every step involved in producing a product or delivering a service. By mapping out both the current and future states of a process, industrial engineers can make informed decisions to optimize workflow, reduce lead times, and improve overall efficiency.

The **current state map** is the first step in VSM, providing a clear visual of the process as it currently exists. Engineers begin by documenting every step, from raw material intake to final product delivery, detailing the flow of materials, actions taken, waiting times, and information exchanges. For example, in an electronics assembly plant, the current state map would include steps like parts assembly, inspection, packaging, and shipping. Each stage of the map shows how materials move and identifies where delays, excess inventory, or redundant steps exist. Data like cycle times, lead times, and inventory levels are included, creating a comprehensive picture of where the process can be improved.

A crucial part of VSM is **identifying value-added versus non-value-added activities**. Value-added activities directly contribute to the final product from the customer's perspective, while non-value-added activities (or waste) do not. For instance, in a painting process, the time spent applying paint is value-added, whereas the time parts wait in a queue is non-value-added. By clearly distinguishing these steps on the map, engineers can prioritize removing or reducing non-value-added activities to create a more streamlined flow.

Once the current state is mapped, engineers develop the **future state map**, envisioning an optimized version of the process. The future state map removes unnecessary steps, reduces wait times, and improves the flow of materials and information. For example, if a manufacturer identifies excessive wait times between machining and assembly due to insufficient scheduling, the future state map might include adjustments to create a continuous flow, reducing these delays. This map serves as a blueprint for lean initiatives, detailing the ideal process configuration to achieve maximum efficiency.

In VSM, **takt time** is an essential metric used to balance production with customer demand. Takt time is calculated by dividing the available work time by customer demand, setting the pace that production should maintain to meet orders without overproduction. For instance, if a factory has eight hours to produce 480 units, the takt time is one unit every minute. By aligning each process step to meet this rate, VSM ensures that resources are used efficiently without creating excess inventory or overburdening any part of the system.

Lead time and cycle time analysis are also integral to VSM. Lead time is the total time from when a product enters the system until it leaves, while cycle time is the time spent working on the product at each stage. VSM helps highlight discrepancies between lead and cycle times, pinpointing areas where products spend excessive time waiting or in transit. For example, if the lead time for an assembly process is 10 hours, but actual cycle time is only 2 hours, this indicates 8 hours of non-value-added waiting. By mapping and analyzing these times, engineers can focus on reducing unnecessary delays, ensuring that the entire process runs more efficiently and responds better to customer demand.

Inventory management is another focus in Value Stream Mapping. Excessive inventory can indicate overproduction or inefficient scheduling, both of which are types of waste in lean manufacturing. On the map, inventory levels are shown at each stage, including work-in-progress (WIP) between stations. For instance, if there is significant inventory buildup between machining and assembly, this suggests a bottleneck or misalignment in production flow. The future state map might recommend implementing a pull system, where each step only produces as needed, ensuring that inventory levels remain aligned with actual demand rather than forecasted estimates.

Kaizen bursts, or rapid improvement opportunities, are often used within VSM to signal specific areas for immediate action. These bursts are marked on the map to indicate places where focused, small-scale changes can significantly impact the flow and eliminate waste. For example, if a workstation frequently experiences delays due to setup time, a kaizen burst might suggest implementing SMED (Single-Minute Exchange of Dies) techniques to reduce that setup time. These targeted improvements are prioritized for rapid implementation, allowing teams to address inefficiencies promptly and make incremental progress toward the future state.

In addition to physical material flow, **information flow** is a critical part of VSM. This includes order information, production scheduling, and feedback mechanisms that coordinate activities across the value stream. Information flow issues, such as delayed communication between departments or inconsistent scheduling, often lead to bottlenecks and misaligned production rates. For instance, if orders are not updated in real-time, production might continue on items with low demand, leading to overproduction. VSM enables engineers to map out information pathways and address communication breakdowns that could disrupt the flow, creating a more responsive and integrated system.

Once the future state map is complete, industrial engineers develop a **roadmap for implementation**. This roadmap includes prioritized action steps, resources needed, and timelines for each phase of improvement. By systematically moving toward the future state, the organization reduces waste incrementally, allowing each change to contribute to a smoother, more efficient process. For example, engineers might start with reducing wait times between major stages, then gradually introduce lean principles like 5S (Sort, Set in order, Shine, Standardize, Sustain) to enhance organization and visual control.

CHAPTER 6: SIX SIGMA AND QUALITY CONTROL

Fundamentals of Six Sigma

Six Sigma is a quality management approach that aims to improve processes by reducing variability and minimizing defects. Developed by Motorola in the 1980s, Six Sigma uses data-driven methodologies and statistical tools to achieve near-perfect quality. The ultimate goal is to reach a defect rate of fewer than 3.4 defects per million opportunities (DPMO), which represents a "six sigma" level of quality —a measure of process capability where 99.99966% of outputs meet specifications.

At the core of Six Sigma is the **DMAIC framework**—Define, Measure, Analyze, Improve, and Control. Each phase provides a structured approach to problem-solving and process improvement.

1. **Define**: In this phase, the project team defines the problem they aim to solve, the goals they want to achieve, and the scope of the project. They also determine the key stakeholders, such as customers and management, and clarify what "success" looks like. For example, if a manufacturing process has inconsistent product dimensions, the Define phase would focus on setting precise goals for dimensional accuracy and identifying how improvements will impact overall quality.

2. **Measure**: The Measure phase involves collecting baseline data on the current process performance. The team identifies key performance indicators (KPIs) and gathers data to quantify the extent of the problem. In the manufacturing example, this might include measuring product dimensions and tracking the frequency of deviations from specification. This phase is essential for establishing a clear, data-backed understanding of the process's current state.

3. **Analyze**: In this phase, the team uses statistical tools to identify the root causes of defects or variations. Techniques like root cause analysis, cause-and-effect diagrams, and hypothesis testing help pinpoint what's driving the variability. For instance, they might find that certain machine settings or environmental factors affect product dimensions. Analyzing data allows the team to isolate which variables impact quality and need improvement.

4. **Improve**: After identifying root causes, the team develops and implements solutions to address them. This may involve adjusting process parameters, redesigning workflows, or training operators. In the Improve phase, solutions are tested, often through pilot programs, to ensure they effectively reduce defects. For example, adjusting a machine's calibration or adding quality checks might reduce dimensional variation in products.

Testing solutions on a smaller scale allows teams to verify results before making widespread changes.

5. **Control**: The Control phase ensures that improvements are sustained over time. The team implements monitoring tools, such as control charts, to track the process and maintain consistency. If deviations begin to appear, these tools can signal a need for corrective action before defects reoccur. Control procedures standardize improvements, making the process more robust and less prone to drift back to previous levels of performance.

In addition to DMAIC, Six Sigma uses **statistical tools** like process capability analysis, regression analysis, and design of experiments (DOE) to understand and improve processes. For example, process capability analysis determines how well a process meets specifications, allowing teams to measure the gap between current performance and desired outcomes. DOE helps identify optimal conditions for a process by testing various combinations of variables, helping to pinpoint the settings that deliver consistent, high-quality results.

Six Sigma also incorporates **roles and certifications** to support its structured approach. Team members are categorized by "belt" levels—Yellow Belt, Green Belt, Black Belt, and Master Black Belt—indicating their expertise and training in Six Sigma. Green Belts typically work on projects part-time, while Black Belts lead projects full-time, applying advanced statistical tools and methodologies. Master Black Belts oversee Six Sigma programs at an organizational level, offering guidance and training.

Another key concept in Six Sigma is the **Critical to Quality (CTQ)** aspect, which focuses on identifying the characteristics of a product or service that are most important to the customer. For example, in a call center, CTQ factors might include wait times, call resolution rates, and agent friendliness. By identifying and focusing on these CTQs, Six Sigma projects align improvements with customer needs, ensuring that quality enhancements have a meaningful impact.

Ultimately, Six Sigma aims to create processes that are both effective and efficient, emphasizing precision and reliability. Through data analysis, structured problem-solving, and continuous monitoring, Six Sigma enables organizations to achieve high-quality outputs and reduce costs by minimizing variability and defects.

Defining and Measuring Quality

In Six Sigma and quality control, defining and measuring quality is essential to establishing consistent, high-performance processes. Quality in Six Sigma is defined by how well a product or service meets customer expectations, often referred to as being **"fit for purpose"**. In practice, this means that quality is not only about meeting technical specifications but also ensuring that outputs meet the specific needs and standards valued by the customer.

The process begins with identifying **Critical to Quality (CTQ)** characteristics. CTQs are the key elements that directly impact the customer's satisfaction with the product or service. For instance, in a smartphone, CTQs might include screen durability, battery life, and processing speed. Each CTQ is determined based on customer expectations and market research, providing a clear focus on what aspects of the product or service need to be maintained or improved. By pinpointing these essential attributes, Six Sigma projects concentrate efforts on areas that make the biggest impact on customer satisfaction and loyalty.

Once CTQs are defined, **quality metrics** are established to measure these attributes consistently. These metrics act as benchmarks for what constitutes acceptable performance. For example, if battery life is a CTQ for a smartphone, a specific target like "10 hours of usage per charge" might be set, providing a measurable standard. These metrics serve as reference points throughout the production process, helping teams track whether the output aligns with desired specifications.

To measure quality effectively, Six Sigma projects rely on **data collection and analysis**. In this phase, engineers collect baseline data on current performance levels, often using tools like control charts or capability analysis. For instance, if dimensional accuracy is crucial in manufacturing, measurements for each batch of products are collected and analyzed to establish current performance levels and detect variation. This data provides a factual basis for understanding the extent of quality deviations and serves as a baseline for future improvements.

Process capability analysis is used to assess how well a process meets quality specifications. This analysis compares the natural variation in a process (measured by standard deviation) with the specification limits for the CTQs. For example, if a product component must be within a specified range, process capability analysis can determine the percentage of units that meet this requirement. If a process has a low capability, it means there's a higher likelihood of producing out-of-spec products, signaling a need for adjustments or improvements.

Defect metrics, such as defects per unit (DPU) or defects per million opportunities (DPMO), are also essential in Six Sigma. These metrics provide a quantitative measure of quality by counting the number of deviations from specifications in a sample of units. For example, if a batch of electronic components has 20 defective items out of 1,000 units, the defect rate is 2%. By calculating DPMO, the team can set clear improvement targets and determine the level of precision needed to reach Six Sigma levels.

In Six Sigma, **Gauge R&R (Repeatability and Reproducibility)** studies validate the accuracy and reliability of the measurement system itself. Gauge R&R ensures that measurement tools and methods consistently capture data with minimal error. If a gauge R&R study reveals a high variation in measurements due to the tools or operators, adjustments are made to reduce inconsistencies. This guarantees that the data collected accurately reflects the process performance, preventing incorrect conclusions and guiding quality improvements based on reliable data.

Process Control and Statistical Quality Control (SQC)

Process control is a central concept in Six Sigma and quality control, ensuring that production processes remain consistent and deliver outputs that meet quality standards. Statistical Quality Control (SQC) is a key technique in process control, using statistical methods to monitor and maintain process performance. Through SQC, Six Sigma practitioners detect variations in real time, enabling quick corrective actions and preventing defects before they impact final output.

One primary tool in SQC is the **control chart**, which visually represents data points over time, allowing engineers to track how a process performs relative to established limits. Control charts have upper and lower control limits that represent the acceptable range of variation. For instance, in a manufacturing process where dimensional accuracy is critical, control charts plot each measurement against these limits. If data points fall within the limits, the process is considered "in control." However, if points fall outside, it signals an issue that requires investigation, such as a machine misalignment or a material inconsistency. Control charts allow for real-time monitoring, alerting teams to shifts or trends that could lead to defects.

Types of control charts are chosen based on the type of data being monitored. For variable data, such as measurements or weights, X-bar and R (range) charts are used. For attribute data, such as counts of defects or nonconforming items, p-charts and c-charts are appropriate. For example, in a process measuring the diameter of a part, an X-bar chart can track the average diameter for each sample, while an R chart tracks the variability within each sample. This dual approach helps maintain both the process mean and variability within acceptable ranges, ensuring consistent quality.

Capability analysis works in tandem with control charts to assess how well a process meets specifications. Capability indices, like Cp and Cpk, measure how closely a process can produce outputs within specification limits. For instance, if Cp and Cpk values are close to 1, it indicates that the process barely meets the specifications. Higher values indicate a process with less variability relative to the specification range, meaning it is more likely to produce outputs within the acceptable range. If capability analysis reveals that the process is not meeting specifications, it signals a need for process improvements, adjustments, or further investigation.

Root cause analysis is another component of SQC, often used when data indicates a process is out of control. Using tools like cause-and-effect diagrams (fishbone diagrams) and the 5 Whys technique, teams investigate the underlying causes of variation. For instance, if control charts show frequent deviations in a machining process, root cause analysis might reveal that tool wear is affecting dimensions. Addressing the root cause, such as implementing a tool maintenance schedule, prevents recurring issues and stabilizes the process.

Six Sigma also emphasizes **Design of Experiments (DOE)** as a way to test and optimize process parameters. DOE systematically changes variables to understand their effects on the outcome, allowing engineers to find optimal settings. For example, in a chemical process, adjusting temperature and pressure simultaneously in a DOE experiment can reveal the best combination for achieving desired product quality. DOE helps control processes by identifying the settings that minimize variation, ensuring consistent results within desired specifications.

Error-proofing (poka-yoke) is another process control technique that uses simple methods to prevent mistakes before they occur. For example, color-coded components or jigs that only fit correctly prevent assembly errors. Error-proofing ensures that even if a defect or deviation could occur, the process design makes it unlikely, reducing the need for costly inspections or rework.

Through SQC and tools like control charts, capability analysis, and DOE, Six Sigma provides a rigorous, data-driven approach to process control. This ensures that processes are not only designed to meet quality standards but are also continuously monitored, allowing for immediate corrections when deviations occur.

Root Cause Analysis and Problem-Solving Tools

In Six Sigma and quality control, Root Cause Analysis (RCA) is essential for understanding and resolving the underlying causes of defects or process issues. Rather than addressing only the symptoms of a problem, RCA digs deep to uncover the actual reasons behind variations or failures in a process. Effective root cause analysis not only corrects issues in the immediate term but also prevents them from recurring, creating a more stable and reliable process. Six Sigma provides a range of problem-solving tools to support RCA, each designed to systematically identify and address these underlying issues.

A primary tool in RCA is the **5 Whys technique**, which involves asking "Why?" multiple times to get to the root of a problem. By continuously probing deeper, engineers can trace the issue back to its origin. For example, if a manufacturing line frequently produces parts with incorrect dimensions, asking "Why?" might reveal that machine settings were adjusted incorrectly. Going further, "Why was it adjusted?" may show that operators didn't have proper training on the calibration process. This method helps uncover layers of contributing factors, ensuring the solution targets the true root cause.

The **cause-and-effect diagram**, or fishbone diagram, is also popular in RCA. This diagram visually maps out all potential causes of a problem, categorizing them into major areas such as People, Process, Equipment, Materials, Environment, and Methods. For example, if the issue is frequent defects in an assembly process, the fishbone diagram might categorize possible causes under headings like Equipment (tool wear or malfunction), People (training issues), and Methods (inconsistent work instructions). By organizing potential causes this way, the team can systematically explore each area, quickly narrowing down the most likely contributors.

Failure Mode and Effects Analysis (FMEA) is another essential Six Sigma tool, particularly useful when tackling complex systems or processes with multiple potential failure points. FMEA involves identifying each possible failure mode—ways in which a process could go wrong—and then assessing the impact, likelihood, and detectability of each failure mode. Each failure mode is assigned a Risk Priority Number (RPN) based on these factors, allowing the team to prioritize actions that address the most significant risks first. For example, in a car assembly line, an FMEA might assess potential failure modes for each part installation step, ensuring high-risk areas receive focused attention. This approach minimizes the chances of defects impacting the final product by proactively addressing the most likely and damaging failure points.

Pareto analysis, based on the Pareto Principle (the 80/20 rule), is used to focus efforts on the few causes that contribute most to a problem. In quality control, Pareto charts visually represent the frequency or severity of different causes, helping teams prioritize. For instance, if a Pareto chart of product defects shows that 80% of issues stem from just two root causes—incorrect part dimensions and assembly errors—teams can focus resources on addressing these areas first. Pareto analysis directs attention to the most impactful factors, allowing for targeted, efficient problem-solving.

Control charts also have a role in RCA by helping detect process variations that might signal a root cause issue. When data points on a control chart fall outside the upper or lower control limits or show a non-random pattern, it indicates an out-of-control condition. For example, if product dimensions fluctuate outside acceptable limits, control charts can reveal when and where these variations occur, prompting investigation into root causes like equipment wear or operator error. By monitoring process stability, control charts allow engineers to proactively identify and address issues before they become significant problems.

Another advanced RCA method in Six Sigma is **Design of Experiments (DOE)**, which helps identify relationships between multiple variables and their impact on output. DOE systematically changes variables to understand their effects, allowing engineers to pinpoint interactions that may not be immediately obvious. For example, if product quality is affected by both temperature and pressure in a chemical process, DOE experiments can determine the optimal settings for each variable to maintain consistent quality. DOE provides statistical evidence to support decisions, reducing trial and error and increasing confidence in the chosen solution.

Kaizen, or continuous improvement events, often incorporate RCA tools to improve processes incrementally. During a Kaizen event, a cross-functional team examines a specific process area, identifying and addressing root causes in a short time frame. For instance, if a packaging line has recurring delays, a Kaizen event might bring together operators, engineers, and supervisors to use 5 Whys and fishbone diagrams to uncover issues in real-time. Kaizen integrates RCA into day-to-day operations, fostering a culture where teams continuously seek and eliminate root causes of inefficiency.

Together, these RCA tools and problem-solving techniques enable Six Sigma teams to systematically identify, analyze, and address the root causes of quality issues. By targeting the true source of defects, RCA strengthens processes, reduces variability, and ensures improvements are lasting and effective.

Implementing Control Charts for Process Monitoring

Control charts are essential in Six Sigma for monitoring process stability and ensuring outputs remain within acceptable quality standards. By displaying data points over time and comparing them to established control limits, control charts provide a visual tool for tracking process variation. This helps identify patterns, trends, or shifts that may indicate an out-of-control process, enabling timely corrective actions.

A standard control chart consists of a **centerline**, which represents the process average, along with upper and lower control limits (UCL and LCL) that define the acceptable range of variation. If data points fall within these limits and show no unusual patterns, the process is considered "in control." However, points outside the limits or non-random patterns suggest a potential issue. For example, if an assembly line produces parts with consistent dimensions, a control chart would show a stable line near the center. If dimensions suddenly shift upwards or downwards, it signals a potential problem, like a machine misalignment, requiring immediate attention.

Different types of control charts are used depending on the data and context. **X-bar and R charts** are commonly used for variable data, such as measurements of weight, length, or temperature. X-bar charts track the average of each sample, while R charts monitor the range within each sample, capturing both the mean and variability. For example, in a machining process where part dimensions are critical, an X-bar chart would track the average diameter, while an R chart would monitor the spread of values. By using both charts together, engineers can ensure that the process remains centered and consistent.

For attribute data, such as counts of defective items or occurrences of specific events, **p-charts** and **c-charts** are effective. A p-chart tracks the proportion of defects within samples, while a c-chart counts the number of defects per unit. In a production line where products are inspected for visual defects, a p-chart might monitor the defect rate in each batch, alerting engineers if the proportion exceeds acceptable limits. Attribute charts are valuable when dealing with pass/fail data, providing a clear way to assess consistency in categorical outcomes.

Implementing control charts requires careful **establishment of control limits** based on historical data or pilot runs, capturing the normal range of variation in a stable process. Engineers typically set the UCL and LCL three standard deviations from the mean, covering 99.7% of expected variation under normal conditions. This approach ensures that most random variation does not trigger false alarms while still detecting significant deviations. If, for example, a process with a mean output of 100 units has a standard deviation of 5 units, control limits would be set at 85 and 115 units, providing a clear threshold for detecting outliers.

Interpreting patterns on control charts is critical for identifying process issues. Patterns like a run of consecutive points above or below the centerline, a trend moving upward or downward, or repeated cycles suggest non-random variation. For instance, a consistent upward trend in part dimensions on an X-bar chart could signal tool wear, requiring maintenance. Detecting these patterns early enables prompt adjustments, preventing issues from escalating.

CHAPTER 7: SUPPLY CHAIN MANAGEMENT

Key Components of Supply Chains

In supply chain management, understanding the key components of a supply chain is essential for designing and managing efficient systems. A supply chain includes all the steps and parties involved in delivering a product or service, from raw material sourcing to delivering finished goods to customers. Each component interacts with others, influencing the entire supply chain's performance in terms of cost, speed, and quality.

The first key component is **suppliers**. Suppliers provide the raw materials, parts, or components necessary to produce a product. The relationship between a company and its suppliers directly impacts the reliability and quality of the end product. Strong supplier partnerships can ensure timely deliveries and consistent quality. For instance, an automotive company relies on steel, glass, and electronics suppliers to provide materials on schedule. To manage these relationships effectively, companies often use **supplier management systems** that track performance, delivery times, and quality levels, allowing them to address issues quickly and maintain a steady flow of materials.

Procurement is the next component, focusing on acquiring goods or services from suppliers at the best price, quality, and terms. Effective procurement involves negotiating contracts, managing supplier relationships, and ensuring that materials meet specifications. For example, a food manufacturer must procure ingredients from multiple suppliers, negotiating prices and terms to stay competitive while maintaining quality standards. Procurement teams balance cost control with quality assurance, using **strategic sourcing** techniques to find suppliers that align with their goals for sustainability, efficiency, and reliability.

Once materials are procured, **production** becomes the central focus. Production refers to the transformation of raw materials into finished goods. This involves manufacturing processes, quality control, and resource allocation. In a computer assembly plant, for instance, production includes assembling components, testing functionality, and packaging finished products. Efficient production relies on **capacity planning** to match production rates with demand and **inventory management** to ensure that parts and materials are available without excess. Production processes must be optimized for consistency and speed, balancing output with quality to meet customer expectations.

Inventory management spans the entire supply chain, from raw materials to finished products. Proper inventory management ensures there are enough materials and products to meet demand without tying up excess capital. In retail, for example, managing seasonal inventory can be challenging as demand fluctuates.

Companies use **inventory control methods** like Just-in-Time (JIT) and Economic Order Quantity (EOQ) to manage stock levels efficiently. JIT aims to reduce inventory by receiving goods only as needed, while EOQ calculates the optimal order quantity that minimizes total costs, including holding and ordering costs. Both approaches prevent overstocking and stockouts, keeping the supply chain responsive to changes in demand.

Another critical component is **warehousing and storage**, where materials and finished goods are held until needed for production or delivery. Warehouses act as storage points that buffer the flow of goods through the supply chain, enabling companies to meet demand without delays. Efficient warehousing involves strategic layout design, automation, and inventory tracking to ensure that products can be retrieved and shipped quickly. In e-commerce, for instance, warehouses are set up to handle high volumes of fast-moving consumer goods, using **automated picking systems** to speed up order fulfillment and reduce manual labor.

Transportation connects each stage of the supply chain, moving raw materials to production facilities and finished products to distribution centers or customers. Transportation includes various modes like trucks, trains, ships, and planes, each with trade-offs in cost, speed, and environmental impact. For instance, air transport is faster but more expensive than ocean shipping, making it suitable for high-value, time-sensitive goods. Logistics teams decide on the optimal transportation methods based on delivery timelines, costs, and customer expectations. Effective transportation management relies on **route optimization** and **load planning** to minimize costs and ensure timely deliveries.

Distribution is the process of delivering finished goods to customers, whether through direct shipping, retail stores, or distribution centers. Distribution channels depend on the company's sales strategy and customer locations. For example, a pharmaceutical company might use specialized distribution centers that handle temperature-controlled storage to meet safety standards. Effective distribution strategies include **demand forecasting** and **distribution network design** to align stock levels with regional demand. Companies often work with third-party logistics providers (3PLs) to expand their reach, improve delivery times, and reduce distribution costs.

A key overarching component is **demand planning and forecasting**. This involves predicting future customer demand to ensure that each stage of the supply chain operates smoothly. Demand planning combines historical data, market trends, and seasonal factors to estimate demand accurately. In retail, demand forecasting helps anticipate high sales periods, like holiday seasons, allowing companies to adjust inventory levels and production schedules in advance. Sophisticated forecasting models use **predictive analytics** to factor in variables like consumer behavior, economic conditions, and competitor actions, helping companies align supply chain activities with expected demand.

Finally, **customer service** is the endpoint of the supply chain, focusing on meeting customer expectations and ensuring satisfaction. Customer service includes order accuracy, on-time delivery, and responsive support for any issues that arise. For instance, in e-commerce, where customers expect fast and accurate deliveries, efficient customer service requires a seamless interaction between warehousing, transportation, and support teams. Customer service data can also feed back into the supply chain, revealing areas for improvement based on common issues or delays. Companies use **customer relationship management (CRM)** systems to track interactions and enhance service quality, which strengthens customer loyalty and provides valuable insights for supply chain adjustments.

Together, these components form a complex network that requires precise coordination. Suppliers, procurement, production, inventory management, warehousing, transportation, distribution, demand forecasting, and customer service each contribute to the overall efficiency and responsiveness of the supply chain. By managing these elements carefully, companies can reduce costs, improve speed, and ensure a reliable flow of goods from origin to customer.

Demand Forecasting and Inventory Control

In supply chain management, demand forecasting and inventory control are essential for maintaining the right balance of stock levels to meet customer demand without overstocking or understocking. Accurate demand forecasting helps predict future customer demand, guiding production, procurement, and inventory decisions. Inventory control ensures that stock levels are managed efficiently, aligning closely with forecasted demand.

Demand forecasting starts by analyzing historical data to identify trends and patterns in customer demand. Seasonal demand fluctuations, for example, are common in retail, where certain items may see high demand during holidays or special events. By examining historical data, supply chain managers can anticipate these spikes and adjust inventory levels accordingly. Advanced forecasting models incorporate multiple factors, such as historical sales, economic indicators, and market trends, providing a more accurate prediction of future demand.

Quantitative forecasting techniques like time-series analysis and regression analysis are frequently used. Time-series analysis examines past sales patterns to project future demand, focusing on seasonal and cyclical trends. For instance, a beverage company might use time-series forecasting to predict increased sales during summer months. Regression analysis, on the other hand, identifies relationships between demand and influencing factors like price changes, advertising, or economic conditions. If a company's sales data shows that demand increases with certain promotions, regression models can help quantify the effect and incorporate it into demand forecasts.

Qualitative forecasting methods are valuable when historical data is limited, such as when launching a new product. Expert judgment, market research, and customer surveys provide insights into expected demand levels. For example, before releasing a new electronic gadget, a company might survey potential customers to gauge interest and estimate demand. Combining qualitative and quantitative methods allows companies to refine their forecasts and reduce the risk of overstocking or understocking.

Once demand forecasts are established, **inventory control** comes into play to manage stock levels efficiently. Inventory control techniques like Economic Order Quantity (EOQ), safety stock calculations, and Just-in-Time (JIT) inventory help optimize the balance between holding costs and stock availability. EOQ is a formula that calculates the optimal order size, minimizing the combined costs of ordering and holding inventory. For example, a company with consistent demand for a product can use EOQ to determine the most cost-effective order quantity and frequency, reducing unnecessary holding costs while ensuring sufficient stock.

Safety stock acts as a buffer to absorb unexpected demand surges or delays in the supply chain. To calculate safety stock, supply chain managers consider the average demand during lead time and potential variations in demand or supply delays. For instance, a retailer with high seasonal variability may carry extra safety stock leading into peak sales periods to ensure it can meet customer demand if demand unexpectedly rises or shipments are delayed. Safety stock minimizes the risk of stockouts, keeping the supply chain responsive and reliable.

Just-in-Time inventory control is a lean strategy that aims to minimize inventory levels by receiving goods only when they are needed in production. JIT requires precise demand forecasts and tight coordination with suppliers to ensure that materials arrive just before they are needed. In the automotive industry, for example, JIT allows manufacturers to receive parts directly to the assembly line rather than holding large amounts of stock, reducing inventory costs and space requirements. However, JIT requires a highly reliable supply chain; any delay or disruption can lead to production stoppages, making accurate demand forecasting even more critical.

Inventory control also involves **ABC analysis**, a method that categorizes inventory based on its value and demand frequency. In ABC analysis, "A" items are high-value or high-demand items that need close monitoring, while "B" and "C" items are lower-value items requiring less frequent oversight. For instance, in a pharmaceutical supply chain, high-demand medications may be classified as "A" items, ensuring that these items are always stocked to prevent shortages. By focusing more on critical items, ABC analysis enables supply chain managers to allocate resources and attention effectively, reducing holding costs for less important stock.

Demand forecasting and inventory control also rely on **inventory tracking technologies** such as barcodes, RFID tags, and inventory management software. RFID tags, for example, enable real-time tracking of products in a warehouse, providing visibility into stock levels, movement, and locations. Integrated inventory management systems synchronize inventory data with demand forecasts, alerting managers when stock reaches reorder points or when excess inventory builds up. This level of visibility helps prevent stockouts, reduce excess, and optimize stock replenishment based on accurate, up-to-date information.

To account for demand fluctuations, supply chain managers may implement **dynamic inventory control**. Dynamic systems adjust stock levels based on current demand trends, sales, and external factors. For example, a retailer facing high demand variability might use a dynamic approach to adjust safety stock levels weekly, accounting for recent sales data and upcoming promotions. This adaptability ensures that inventory remains aligned with real-time demand, reducing costs associated with excess inventory or emergency orders.

Ultimately, effective demand forecasting and inventory control work together to ensure the supply chain is responsive, cost-efficient, and capable of meeting customer expectations. By leveraging advanced forecasting models, precise inventory management strategies, and real-time tracking, companies maintain a competitive edge and improve operational efficiency in complex, demand-driven markets.

Supplier and Customer Relationship Management

Supplier and customer relationship management are critical to a well-functioning supply chain, as strong relationships with both suppliers and customers lead to increased reliability, responsiveness, and satisfaction throughout the supply chain. Effective relationship management involves collaboration, communication, and alignment of goals with both upstream (suppliers) and downstream (customers) partners.

In **supplier relationship management**, companies focus on building reliable and collaborative partnerships with suppliers. A strong supplier relationship ensures timely delivery of materials, consistent quality, and flexibility to respond to demand changes. For example, a company producing consumer electronics relies on suppliers for essential components like microchips and display screens. Through consistent communication and regular performance reviews, the company can ensure suppliers meet quality standards and delivery schedules. In some cases, companies use **supplier scorecards** to evaluate performance based on criteria like quality, cost, delivery, and service. This structured approach promotes transparency and encourages continuous improvement from suppliers.

Strategic sourcing and long-term contracts with key suppliers strengthen supplier relationships, allowing companies to lock in favorable pricing and secure supply reliability. In industries like automotive, where precision parts are essential, companies often form long-term partnerships with a select group of suppliers who understand their specific requirements. This partnership approach reduces risk and improves supply chain stability by ensuring that suppliers are committed to the company's standards and timelines. Companies may also invest in **supplier development programs**, helping suppliers improve processes or adopt new technologies to enhance quality and efficiency.

Communication is critical in supplier relationship management. Regular updates on demand forecasts, production schedules, and quality requirements help suppliers prepare and adapt to changes. For example, a food manufacturer might communicate forecasted increases in demand for a particular ingredient during a holiday season, allowing suppliers to adjust production and avoid shortages. This level of collaboration helps suppliers meet demand fluctuations more effectively, preventing disruptions and maintaining a steady flow of materials.

Customer relationship management (CRM) focuses on understanding and meeting customer needs, ensuring that products or services meet expectations for quality, delivery, and support. Strong customer relationships drive customer loyalty, repeat business, and valuable insights into market trends. In retail, for instance, CRM systems track customer purchase histories, preferences, and feedback, helping companies tailor products and services to meet specific customer segments.

Effective CRM relies on gathering and analyzing customer data to make informed decisions. For example, a company that sells electronics may use CRM data to track which product features customers value most, allowing for product design improvements or targeted promotions. By understanding customer preferences, companies can align inventory, production, and distribution strategies to meet real demand, enhancing customer satisfaction and reducing excess stock.

CRM also involves **post-sales support and service**, which are essential in industries where product reliability and after-sales care influence customer loyalty. In sectors like healthcare or automotive, companies maintain support teams to handle customer inquiries, provide maintenance assistance, and address issues quickly. Good post-sales support improves customer trust and satisfaction, and it often generates useful feedback for future product improvements.

Both supplier and customer relationship management benefit from technology. **Supplier relationship management software** helps track supplier performance, communication, and compliance, while **CRM systems** centralize customer data, providing insights that help align supply chain activities with customer demand. These technologies make it easier to analyze data, forecast trends, and respond to partner needs effectively.

Logistics and Distribution Planning

In supply chain management, logistics and distribution planning are critical for moving products efficiently from production facilities to customers. Logistics encompasses the movement, storage, and flow of goods, while distribution planning focuses on selecting the most effective strategies to deliver products to end users. Both functions aim to optimize delivery speed, reduce costs, and ensure reliability in meeting customer demand.

Transportation management is a key component of logistics, as it determines the methods, routes, and scheduling of shipments. Selecting the appropriate transportation mode—road, rail, air, or sea—depends on factors like delivery speed, cost, and product type. For instance, perishable items like fresh produce require rapid delivery to maintain quality, often using air or refrigerated trucking. Bulk materials, on the other hand, are typically transported via rail or sea due to lower costs and reduced urgency. Transportation management also involves optimizing routes to reduce transit times, fuel consumption, and costs. Route optimization tools consider factors like distance, traffic, and delivery windows, helping logistics managers minimize delays and enhance delivery efficiency.

Warehouse and inventory management are central to effective distribution. Warehouses act as storage points that enable a consistent flow of goods through the supply chain. Proper warehouse management includes organizing inventory to minimize retrieval time, tracking stock levels to prevent stockouts, and arranging storage layouts to streamline picking and packing processes. For instance, in an e-commerce warehouse, fast-moving items are often placed near packing stations to speed up fulfillment, while slower-moving products are stored in less accessible areas. Inventory management systems, such as Warehouse Management Systems (WMS), provide real-time stock visibility, allowing logistics teams to monitor inventory levels, track inbound and outbound movements, and replenish stock as needed to maintain a smooth flow.

Cross-docking is a distribution technique that reduces the need for long-term storage by moving incoming goods directly to outbound transportation. In a cross-docking facility, products arrive at inbound docks, are sorted, and then loaded onto outbound vehicles based on delivery destinations. For instance, a retailer may use cross-docking to receive products from multiple suppliers, sort them by store destination, and dispatch them quickly to avoid warehouse holding costs. Cross-docking shortens lead times, cuts storage expenses, and improves delivery speed, making it especially effective for high-turnover products or time-sensitive shipments.

Order fulfillment is the process of completing customer orders accurately and efficiently, encompassing picking, packing, and shipping. Distribution planning focuses on optimizing these tasks to meet customer expectations for fast and reliable delivery. Picking involves locating and retrieving items for an order, while packing ensures products are securely prepared for transit. Shipping then involves coordinating transportation to meet delivery deadlines. E-commerce companies, for instance, rely on efficient order fulfillment systems to handle high volumes of small orders, using technologies like barcode scanning, automated sorting, and real-time order tracking to streamline the process. Efficient fulfillment reduces lead times and enhances customer satisfaction.

In distribution planning, **network design** determines the optimal locations for warehouses, distribution centers, and transportation routes. A well-designed network minimizes transportation costs and lead times by positioning facilities strategically based on customer locations and demand patterns. For example, a company with a nationwide customer base may use a hub-and-spoke network, positioning central distribution centers near large markets to serve regional customers. Network design also involves analyzing transportation costs, delivery times, and inventory requirements to find the right balance between centralized and decentralized storage. Decentralized networks offer faster delivery to regional customers but increase warehouse and inventory costs, while centralized networks reduce storage expenses but may lead to longer delivery times.

Demand forecasting is integrated into logistics and distribution planning to align inventory and delivery schedules with expected demand. By predicting customer needs, logistics managers can ensure sufficient stock at distribution points, avoiding stockouts and minimizing excess inventory. In retail, for instance, seasonal demand forecasting helps prepare for peak periods like holidays, adjusting stock levels and transportation capacity to handle increased volume. Accurate demand forecasts reduce the risk of overstocking or understocking, enabling smoother distribution and cost efficiency.

Reverse logistics handles the return flow of goods, such as returns from customers or unsold stock. Efficient reverse logistics ensures that returned items are quickly processed, restocked, or recycled, reducing costs and minimizing waste. In e-commerce, where returns are common, an effective reverse logistics process might involve automated systems to categorize returned items based on condition, determining whether they can be restocked or require disposal. Managing returns efficiently improves customer experience and reduces inventory losses.

Technology integration supports logistics and distribution planning by providing data visibility, tracking capabilities, and real-time insights. Transportation Management Systems (TMS) coordinate shipments, optimize routes, and track deliveries, while WMS manage inventory within warehouses, ensuring accurate stock levels and efficient order picking. For example, a TMS can automatically assign shipments to carriers, track real-time locations, and monitor delivery status. WMS, on the other hand, helps warehouse teams locate items, update inventory records, and manage replenishment schedules. Integrated technology allows supply chain managers to make data-driven decisions, react to disruptions, and optimize the entire logistics network.

Last-mile delivery—the final step of getting products to customers—is often the most complex and costly segment of logistics. In urban areas, companies might use bike couriers or electric vehicles to navigate traffic and reduce environmental impact, while in rural areas, drone deliveries or third-party logistics providers might be more effective. Last-mile delivery services aim to meet customer expectations for speed, convenience, and transparency. Innovations in last-mile delivery, such as lockers and local pick-up points, help manage costs and offer customers more options for receiving their orders.

CHAPTER 8: OPERATIONS RESEARCH AND OPTIMIZATION TECHNIQUES

Linear Programming and Optimization Models

Linear programming (LP) is a mathematical technique used in operations research to optimize decision-making when resources are limited. In industrial engineering, LP helps determine the best way to allocate resources—like materials, labor, and time—to maximize outputs or minimize costs. Optimization models provide structured methods to solve these complex problems by defining clear objectives and constraints, allowing engineers to make informed decisions in various applications, from production scheduling to transportation planning.

In linear programming, the goal is to optimize an **objective function**, which represents the outcome that needs improvement, such as profit maximization or cost minimization. The objective function is usually a linear equation, formulated with decision variables that represent quantities to be determined. For example, a factory may want to maximize profit by deciding how many units of two different products to produce. Here, the objective function could be maximizing total profit, calculated by summing the profit per unit of each product multiplied by the number of units produced.

LP models also include **constraints**—linear inequalities that represent the limitations within the system, such as material availability, labor hours, or machine capacity. Constraints define the feasible region, which is the set of all possible solutions that satisfy these limitations. For instance, if a factory has limited hours of machine time, a constraint would limit the number of units produced by the total hours available. Constraints ensure that the solution respects real-world limits, preventing outcomes that are theoretically possible but practically impossible.

To solve LP problems, linear programming uses a method called the **Simplex algorithm**. This algorithm starts at a corner of the feasible region and moves along the edges, checking each vertex until it finds the optimal solution—either the highest or lowest point on the objective function. The Simplex method is efficient for large problems and guarantees finding an optimal solution if one exists. For example, in a transportation network where the objective is to minimize cost, the Simplex algorithm explores different routes and allocations until it identifies the lowest-cost solution that satisfies all capacity and demand constraints.

One common application of linear programming is in **production planning**. Manufacturers use LP models to determine the optimal production levels for multiple products, taking into account constraints like resource availability and demand. In a factory that produces furniture, for instance, LP can help decide how many tables and chairs to make, given limited wood and labor hours, to maximize profit. The objective function here might be maximizing revenue, while constraints include material quantities, labor availability, and storage space. By solving the LP model, the factory can allocate resources efficiently to maximize profitability.

Another critical application of LP is in **transportation and logistics**. LP models can minimize transportation costs by determining the optimal routes and quantities to ship between suppliers, warehouses, and retailers. In a supply chain for a retailer with multiple warehouses and stores, an LP model can calculate the most cost-effective way to distribute products while meeting demand at each location. Constraints like transportation capacity, demand requirements, and delivery timelines shape the solution. This type of optimization allows companies to reduce costs and improve service levels by ensuring products reach their destinations through the most efficient routes.

Dietary and blending problems are also solved using linear programming. These problems involve finding an optimal mix of ingredients or materials to meet certain nutritional or quality requirements at the lowest cost. In agriculture, for instance, LP models can help farmers determine the ideal feed mix for livestock that meets nutritional needs while minimizing expenses. Here, the objective function could be minimizing cost, while constraints ensure that the feed provides required levels of protein, fat, and other nutrients. Blending problems are also common in industries like oil refining, where LP helps decide the best mix of raw materials to meet product specifications.

In addition to the Simplex algorithm, **integer programming** and **mixed-integer programming (MIP)** are extensions of linear programming. While LP assumes decision variables can take any continuous values, integer programming requires variables to be integers, which is essential in cases where only whole units make sense, such as the number of trucks or machines. MIP combines both continuous and integer variables, providing more flexibility. For example, in workforce scheduling, where a company may need a whole number of shifts, MIP models enable optimal scheduling while accounting for discrete resources.

Sensitivity analysis is a valuable tool within LP to assess how changes in input data, like costs or available resources, impact the optimal solution. This analysis helps industrial engineers understand the robustness of their decisions and prepares them for possible variations in real-world scenarios. For instance, in a factory that produces chemicals, if the cost of a raw material increases, sensitivity analysis shows whether the current production plan remains feasible or requires adjustment. This insight is particularly useful in volatile markets, where prices and availability can change frequently.

Goal programming is another extension of LP, used when multiple objectives need to be balanced. Unlike standard LP, which optimizes a single objective, goal programming can handle multiple, sometimes conflicting, objectives. For instance, a company might want to minimize costs while also maximizing customer satisfaction by reducing delivery times. Goal programming assigns priority levels to each objective and finds a solution that best satisfies these priorities. This approach is especially helpful in complex systems where trade-offs between different goals are necessary.

Linear programming and optimization models allow industrial engineers to make data-driven, optimal decisions across a variety of applications. By setting up objective functions, incorporating real-world constraints, and using algorithms like the Simplex method, LP models provide a clear, structured way to navigate complex decisions and achieve the best possible outcomes.

Decision Trees and Decision-Making Tools

Decision trees are used in operations research for analyzing and visualizing choices in complex decision-making processes. They represent potential decisions and their outcomes in a tree-like structure, allowing managers and engineers to evaluate various scenarios, costs, benefits, and risks. Each **branch of a decision tree** represents a possible action or decision, while subsequent branches indicate outcomes, including probabilities and payoffs. Decision trees enable a step-by-step analysis of multiple paths, providing clarity in making the best decision based on probable outcomes and associated risks.

The **decision tree structure** starts with a root decision node, which branches out into possible choices. Each choice leads to further nodes that represent possible events or outcomes, creating additional branches. For example, in choosing a new supplier, a company could face decision branches with potential outcomes, such as reliability, cost, and lead times. At each decision point, the tree branches out, presenting outcomes like delays or cost savings, allowing decision-makers to visualize and evaluate the consequences of each path.

Decision trees incorporate **probabilities** and **payoffs** for each possible outcome, making them effective for analyzing uncertain scenarios. Probabilities, often based on historical data or expert estimates, are assigned to each possible event, and payoffs represent the gains or losses associated with each outcome. For instance, a logistics company choosing between two shipping routes can assign probabilities to the likelihood of delays or cost increases on each route and estimate the financial impact. By calculating the expected value of each decision path (multiplying the probability by the payoff), decision-makers can select the option with the highest expected benefit or lowest cost.

Expected Value (EV) is a common metric in decision tree analysis, calculated by summing the expected payoffs across all potential outcomes for each decision path. If a manufacturing plant needs to decide between investing in a new machine or repairing an existing one, it can assign probabilities and costs to outcomes like successful operation, machine failure, or maintenance requirements. By calculating the EV for each decision, the plant management can determine which option yields the best financial outcome based on risk and potential reward.

Decision trees are also effective in **sensitivity analysis**, allowing decision-makers to see how changes in probabilities or payoffs affect outcomes. By adjusting probabilities for uncertain factors, such as supply disruptions or fluctuating demand, analysts can assess the impact of different scenarios on overall decision quality. For example, if a retailer is considering expanding its product line but faces uncertain demand, sensitivity analysis with decision trees helps visualize how different demand levels could impact profits or losses. This flexibility makes decision trees highly valuable in dynamic, uncertain environments.

In addition to decision trees, other decision-making tools support complex analysis. **Payoff matrices** provide a grid-like representation of potential decisions and their outcomes, listing the payoffs for each combination of choices and events. For example, a company deciding between three different production methods can use a payoff matrix to evaluate cost, quality, and production speed for each method. The matrix format makes it easier to compare outcomes across multiple criteria, simplifying the decision-making process when several factors must be weighed simultaneously.

Multi-criteria decision analysis (MCDA) is another method that considers multiple factors to make balanced decisions. MCDA assigns weights to different criteria based on their importance, helping decision-makers evaluate options that involve trade-offs between objectives. For instance, when selecting a supplier, a company might weigh criteria like cost, reliability, and lead time. By scoring and weighting each criterion, MCDA helps determine the best supplier choice, even when one option isn't clearly superior across all criteria.

Risk analysis tools complement decision trees by evaluating the risks associated with each potential choice. Tools like Monte Carlo simulations generate a range of possible outcomes based on different scenarios, assigning probabilities to each outcome. In capital investment, for example, Monte Carlo simulation can model how different economic conditions affect project returns, helping managers make risk-informed decisions. This probabilistic approach is valuable when decisions involve significant uncertainty or complex variables.

Decision trees and associated tools help organizations make well-informed, data-driven decisions by breaking down complex choices into manageable steps. With visual representations, sensitivity analysis, and integrated risk assessments, decision trees provide clarity in navigating uncertainty and optimizing outcomes.

Simulation Modeling in Process Optimization

Simulation modeling is a popular framework in operations research for optimizing complex processes. By creating a virtual representation of a real-world process, simulation allows engineers and decision-makers to test different scenarios, measure outcomes, and predict how changes will affect performance. Simulation models use a combination of variables, probabilities, and constraints to mimic the interactions within a system, enabling organizations to make data-driven decisions without disrupting actual operations.

Discrete-event simulation (DES) is commonly used to model processes where events occur at specific points in time, such as manufacturing, logistics, and customer service. DES focuses on individual events that trigger changes within the system, like the arrival of materials in a warehouse or the completion of a machine cycle. For example, in a factory assembly line, DES can simulate each stage of production, capturing delays, bottlenecks, and idle times. By testing different production schedules or machine configurations, engineers can identify inefficiencies and improve throughput without affecting the physical assembly line.

In logistics, DES helps model warehouse operations, allowing companies to optimize layout, storage strategies, and picking routes. For instance, a distribution center might use DES to simulate peak-season demand, testing different staffing levels, inventory placement, and packing methods. These simulations reveal which configurations maximize order fulfillment speed and minimize handling costs. Additionally, DES captures the variability in demand or lead times, helping logistics managers prepare for disruptions.

Continuous simulation models are useful for systems where changes occur in a fluid, uninterrupted manner, such as chemical processing or thermal dynamics. In continuous simulation, variables change continuously over time rather than in discrete steps, making it suitable for processes that require fine-tuned adjustments. In an oil refinery, for example, continuous simulation can model the flow rates and temperature changes across processing units, helping engineers optimize settings to maximize yield and reduce waste.

Monte Carlo simulation is widely used in risk analysis and decision-making to assess the impact of uncertainty on outcomes. Monte Carlo models rely on random sampling to explore a range of possible scenarios, assigning probabilities to each outcome. In financial forecasting, for example, a Monte Carlo simulation might evaluate the impact of different market conditions on investment returns. By running thousands of iterations, the simulation reveals the likelihood of various results, allowing decision-makers to assess risks and make informed choices.

Simulation is also crucial for **queuing analysis**, where the goal is to optimize waiting times and service levels. In customer service centers, hospitals, and retail stores, queuing simulations help determine the optimal number of servers, checkouts, or staff to handle varying levels of demand. For instance, in a hospital emergency room, a simulation model can test different staffing levels and triage protocols to reduce patient wait times and improve care quality. By identifying bottlenecks and optimizing resource allocation, queuing simulations enhance service efficiency and customer satisfaction.

In **inventory management**, simulation models test different stock levels, reorder points, and lead times to minimize holding costs while ensuring product availability. For instance, in an automotive parts warehouse, a simulation can evaluate how different inventory strategies impact stockouts and order fulfillment times. By modeling seasonal demand, lead time variability, and reorder quantities, simulation helps supply chain managers balance costs with service levels, ensuring optimal inventory performance.

Agent-based modeling (ABM) offers another approach to simulation by representing individual agents, such as machines, employees, or customers, and their interactions within a system. ABM is most useful in systems where the behavior of individuals significantly influences overall performance. In retail, for example, an ABM simulation could model customer movement within a store, testing different layouts to improve traffic flow and maximize sales. From observing how individual agents react to layout changes or promotions, ABM provides insights into customer behavior and helps optimize store design.

Simulation modeling is also valuable in **project management**, where it predicts project timelines and resource needs under various scenarios. Techniques like **project network simulation** model the dependencies between tasks, capturing potential delays and resource constraints. In a construction project, for example, simulation can model the impact of weather delays, material shortages, or workforce availability, helping project managers create realistic schedules and contingency plans.

One of the strengths of simulation is its ability to support **what-if analysis**, allowing engineers to test potential changes in a virtual environment. In manufacturing, what-if scenarios could include introducing a new machine, increasing production rates, or adjusting shift schedules. Simulation helps predict the outcomes of these changes without committing resources, providing insights into whether the changes will lead to improved performance or unintended bottlenecks.

Data integration with simulation tools further enhances their accuracy and relevance. By feeding real-time data from sensors, production lines, or ERP systems into the simulation, organizations can update models to reflect current conditions. In predictive maintenance, for example, a simulation model might use live machine data to predict failure times and optimize maintenance schedules. This proactive approach reduces downtime and extends equipment life, keeping the process running smoothly.

Simulation models also facilitate **optimization** through repeated testing of different configurations. By running multiple scenarios and comparing results, organizations can identify the most efficient setup for a process. In a call center, for instance, simulation can test various staffing and scheduling options to meet service-level targets with minimal labor costs. Simulation-driven optimization provides clear recommendations based on evidence, guiding decisions that improve operational efficiency.

Simulation modeling offers a dynamic, data-rich approach to process optimization, allowing organizations to make informed decisions that enhance performance, efficiency, and reliability. Through tools like DES, Monte Carlo simulations, and ABM, simulation provides a versatile framework for addressing complex challenges in a controlled, risk-free environment.

CHAPTER 9: ERGONOMICS AND HUMAN FACTORS

Understanding Human Capabilities and Limitations

Understanding human capabilities and limitations is central to designing work environments that enhance safety, comfort, and efficiency. Ergonomics and human factors engineering study these capabilities and limitations to align work tasks, tools, and environments with the physical and cognitive abilities of workers, reducing fatigue and the risk of injury while boosting productivity.

Physical capabilities vary widely among individuals, influenced by factors like age, strength, and flexibility. When designing workstations, engineers consider the range of motion and strength needed for tasks. For example, lifting heavy objects repeatedly can strain muscles and joints, particularly in the lower back. To minimize this, ergonomic designs often include adjustable lift tables, mechanized assist devices, or height-adjustable work surfaces, which reduce the need for excessive bending or reaching. Understanding these physical limits helps create spaces where workers can perform tasks within a safe, efficient range, reducing muscle strain and cumulative injuries.

Visual capabilities also vary and are crucial in tasks that require attention to detail. Eyesight changes with age, affecting clarity, depth perception, and sensitivity to light. In a detailed assembly process, for instance, poor lighting or small text on labels can slow down work and lead to errors. Designing with visual limitations in mind means using adequate lighting, placing monitors at comfortable viewing distances, and using larger, high-contrast text or symbols. Adjustments like these reduce eye strain and allow workers to perform visually demanding tasks with greater accuracy and less fatigue.

Hearing capabilities are another consideration, especially in noisy environments like factories. Exposure to high noise levels can not only impair hearing over time but also create stress, decrease focus, and complicate communication. Understanding auditory limitations, engineers can design systems with quieter machinery, sound-dampening materials, or noise-canceling headsets. In some cases, using visual signals or vibration cues can effectively replace audible alarms in environments where hearing is limited. By designing workplaces to accommodate different levels of hearing, engineers improve safety and communication, even in high-noise areas.

Cognitive limitations impact how workers process information, make decisions, and respond to changes in their environment. Cognitive load varies by task complexity, the number of decisions required, and the presence of distractions. For instance, in a control room with multiple screens and ongoing alarms, operators may become overwhelmed and prone to errors. To manage cognitive load, human factors engineers simplify information presentation, group related controls, and provide clear, unambiguous labels. Minimizing unnecessary information and organizing tasks logically helps workers focus, improving accuracy and reducing errors in demanding environments.

Reaction time is another factor that varies between individuals and can be influenced by task complexity, fatigue, and environmental conditions. In roles where fast response is critical, like operating heavy machinery or driving, ergonomics involves minimizing distractions and ensuring controls are easy to reach and operate. For example, placing emergency stop buttons within arm's reach and ensuring they are clearly visible allows for quick, instinctive action. Designing with reaction time in mind reduces the likelihood of accidents and supports worker safety in high-stakes scenarios.

Mental and physical endurance are also key in ergonomic design. Tasks requiring sustained attention or repetitive motions over long periods can lead to mental fatigue, physical exhaustion, and a decline in performance. Engineers account for endurance limits by designing tasks that allow regular breaks, varying activities to prevent repetitive strain, and providing supportive seating or anti-fatigue mats. For example, in a production line, rotating workers through different tasks reduces physical stress on specific muscle groups and keeps attention levels high, preventing mental burnout.

Anthropometric data—measurements of human body dimensions—guide the design of workstations, tools, and controls to fit a wide range of body sizes and shapes. For example, adjustable chairs and desks accommodate variations in height, arm reach, and leg length, providing each worker with a comfortable, efficient workspace. This data also helps design tools that fit naturally in the hand, reducing strain on fingers, wrists, and arms. By designing based on human dimensions, engineers ensure that equipment fits users naturally, promoting comfort and reducing awkward postures.

Temperature tolerance is another consideration, particularly in environments where extreme heat or cold affects performance. Prolonged exposure to extreme temperatures can lead to fatigue, dehydration, or even health risks. In high-heat environments like foundries, engineers might design ventilation systems, provide cooling vests, or schedule frequent breaks. In colder environments, they might incorporate heaters, insulated workstations, or appropriate clothing recommendations. Managing temperature in workspaces supports physical endurance, safety, and comfort.

Workstation Design and Ergonomic Assessment

Workstation design is a core element of ergonomics, focused on creating efficient, comfortable spaces that align with the human body's natural posture and movement. By carefully designing workstations, engineers can minimize repetitive strain, reduce discomfort, and improve productivity. **Adjustability** is a key factor in ergonomic workstation design, as it allows the workstation to accommodate individuals of various heights and body types. Adjustable desks, for instance, allow workers to switch between sitting and standing, which can reduce strain on the lower back and legs. Similarly, chairs with adjustable height, lumbar support, and armrests help align the spine naturally, preventing back and shoulder pain.

Work surface height is also crucial. For seated tasks that require precision, such as assembly or inspection, the work surface should be positioned just above elbow height. This height minimizes the need for reaching or bending, keeping the arms at a comfortable level that reduces strain on the shoulders and wrists. Conversely, for tasks that require more force, like packing or heavy lifting, the surface should be lower, allowing for better leverage. Engineers often conduct ergonomic assessments to determine optimal surface heights, ensuring that workstations match the demands of the tasks performed.

Tool and material placement impacts ease of access and productivity. Frequently used items should be placed within the primary reach zone—close enough to grab without stretching. For example, in a machine operation workstation, tools and controls are often placed within arm's reach, preventing excessive reaching or bending. Parts bins, tool racks, and control panels are designed to keep essential items within a worker's reach, eliminating unnecessary movement and saving time. Items that are rarely used can be placed farther away, minimizing clutter and improving workflow efficiency.

Anti-fatigue mats are essential in standing workstations, as they provide a cushioned surface that reduces pressure on the legs and lower back. Hard floors can lead to fatigue and joint pain over time, while anti-fatigue mats encourage subtle foot movements that promote blood flow. For tasks that require standing, such as assembly lines or packing stations, these mats significantly improve comfort and reduce fatigue-related injuries.

Lighting is another critical element of ergonomic workstation design, especially in tasks that require visual precision. Proper lighting reduces eye strain and helps workers focus on detailed tasks without overworking their vision. In workstations where screens are involved, adjustable lighting can prevent glare, while task lighting is essential for assembly work to enhance visibility. Engineers assess lighting needs based on task requirements, ensuring brightness levels are appropriate to reduce strain and improve focus.

Ergonomic assessment involves evaluating workstation setup, tools, and worker posture to identify potential issues. Assessments may involve observing workers' movements and identifying repetitive actions or awkward postures. For instance, if a worker frequently twists or bends, the assessment might recommend repositioning materials or tools to create a more natural workflow. This feedback loop between design and assessment ensures that workstations evolve to meet ergonomic standards effectively, aligning design with real-world use.

Through ergonomic assessment, engineers can recommend adjustments or redesigns to improve workstation design continuously. These changes support a healthy, efficient workspace, minimizing risks and enhancing long-term worker comfort.

Cognitive Ergonomics and Usability

Cognitive ergonomics focuses on designing systems and work environments that align with the mental capabilities of workers, optimizing ease of use and reducing cognitive load. Usability is a primary concern, as it ensures that systems, tools, and interfaces are intuitive and efficient for the user, reducing errors and improving performance.

Information display is critical in cognitive ergonomics, especially in environments with complex controls or extensive data, such as control rooms or dashboards. When designing information displays, engineers prioritize clarity, organizing information hierarchically and grouping related data. For example, in an aircraft cockpit, critical information—like altitude, speed, and fuel levels—appears in central, easy-to-read locations, while less critical data is displayed peripherally. This layout minimizes cognitive overload and allows operators to make quick, accurate decisions under pressure by focusing only on the most relevant information.

Interface design is another key element of cognitive ergonomics, as intuitive interfaces reduce the mental effort required to interact with complex systems. Controls should be clearly labeled, and the layout should flow logically, aligning with the operator's natural movements. In a manufacturing control panel, for example, start and stop buttons are often differentiated by color and placement, making them easy to locate and use quickly. Usability testing ensures that the design supports efficient workflows and prevents mistakes that arise from unclear controls or confusing layouts.

Consistency and predictability in system responses enhance usability by allowing operators to develop expectations for how the system will react. For instance, if pressing a button triggers an immediate feedback response—such as a light indicating the machine is active—the operator builds a sense of trust and familiarity with the system. Inconsistencies, like delayed responses or unexpected behaviors, increase cognitive load and can lead to mistakes. Engineers ensure that system responses are predictable, allowing workers to develop routines that improve efficiency and reduce stress.

Cognitive load management is essential, particularly in multitasking environments like air traffic control or emergency response centers, where operators process large amounts of information quickly. Cognitive ergonomics aims to reduce unnecessary complexity by limiting information to what is immediately relevant. For example, filtering out non-critical alarms in a medical monitoring system prevents distraction, allowing healthcare staff to focus on essential data. Cognitive ergonomics emphasizes simplicity, ensuring workers aren't overwhelmed by excess information, which can lead to fatigue and errors.

Error reduction is a central goal of cognitive ergonomics, achieved by designing systems that support decision-making and reduce the likelihood of mistakes. For instance, error-proofing features, like confirmations for critical actions, help operators avoid accidental inputs. In industrial settings, machines may require a two-step activation process for safety-critical operations, ensuring that actions are deliberate and preventing accidental activation. By anticipating potential errors, cognitive ergonomic design improves safety and reliability in complex systems.

Through usability and cognitive ergonomics, engineers create environments and tools that align with human mental processes, simplifying tasks and reducing strain. By focusing on clear information display, consistent interfaces, and manageable cognitive loads, cognitive ergonomics enhances user experience, safety, and productivity.

Health and Safety in the Workplace

Health and safety in the workplace are fundamental aspects of ergonomics, aiming to protect workers from injury, illness, and stress. A well-designed workplace supports not only physical health but also mental well-being, fostering an environment where workers feel safe and productive.

Physical hazards are a primary concern in ergonomic design. Engineers assess potential risks such as repetitive strain, poor posture, and excessive lifting, which can lead to musculoskeletal injuries over time. For instance, in a warehouse, lifting heavy items repeatedly can strain the lower back. Ergonomic assessments identify such risks and suggest solutions, like using lifting equipment, providing adjustable workstations, or redesigning workflows to minimize strain. These measures prevent chronic injuries, ensuring a healthier workforce.

Environmental factors like lighting, temperature, and noise impact workers' comfort and concentration. Poor lighting strains vision, while excessive noise levels contribute to stress and hearing damage. Temperature extremes, whether too hot or cold, can also cause fatigue and discomfort. Engineers address these issues by designing appropriate lighting, using soundproofing materials, and installing HVAC systems to regulate temperature. Comfortable environmental conditions improve focus and reduce the likelihood of accidents caused by fatigue or distraction.

Mental health considerations are also part of workplace safety, particularly in high-stress jobs where employees face cognitive overload. Repetitive tasks, intense deadlines, or constant monitoring can lead to burnout and anxiety. To reduce mental strain, ergonomic designs often incorporate break schedules, task variation, and clear role definitions, giving employees the structure and downtime they need to recharge. For instance, in control rooms, regular breaks are essential to maintain focus and decision-making accuracy. Such adjustments support mental well-being, reducing stress-related incidents.

Emergency preparedness involves designing workplaces that facilitate quick, safe responses to accidents or hazards. Clear signage, accessible exits, and emergency equipment placement are critical. For example, fire extinguishers and first-aid kits should be clearly marked and readily available, especially in areas with potential fire risks or heavy machinery. Training employees on emergency protocols further ensures they can respond calmly and effectively during incidents.

Ergonomic safety measures create a proactive approach to health and safety, identifying risks before they result in injuries. This approach prioritizes the well-being of employees and promotes a safer, more productive workplace.

CHAPTER 10: PROJECT MANAGEMENT FOR INDUSTRIAL ENGINEERS

Project Life Cycle and Phases

In project management, the **project life cycle** provides a structured framework that guides projects from start to finish. Each phase in the life cycle—Initiation, Planning, Execution, Monitoring, and Closing—has distinct tasks and goals that help keep the project on track, ensuring it meets deadlines, budget constraints, and quality requirements. Industrial engineers use the project life cycle to manage resources, minimize risks, and optimize outcomes efficiently.

The **Initiation phase** marks the formal beginning of a project. In this phase, project objectives are defined, feasibility is assessed, and a project charter is created. The project charter is a document that outlines the project's purpose, scope, high-level requirements, stakeholders, and objectives. For instance, if an industrial engineer is managing the installation of new equipment in a manufacturing plant, the initiation phase would involve determining whether the equipment aligns with the company's production goals and reviewing cost and resource requirements. By the end of this phase, decision-makers evaluate the project's viability based on its goals and projected benefits, approving or rejecting the project before moving forward.

In the **Planning phase**, detailed project plans are developed. This phase includes creating a work breakdown structure (WBS), setting timelines, defining tasks, and allocating resources. The WBS breaks down the project into smaller, manageable tasks, making it easier to organize and track progress. For example, in a facility expansion project, the WBS might separate tasks into categories like site preparation, equipment installation, and system integration, with each broken down further. The planning phase also involves risk assessment, where potential issues are identified, and mitigation strategies are established. Risk planning helps industrial engineers anticipate disruptions and develop contingency plans, ensuring the project can adapt to unexpected challenges without significant delays.

Setting a **project schedule** is a critical part of planning. Schedules are created using techniques like the Critical Path Method (CPM) or Gantt charts to identify task sequences and dependencies. For instance, if an industrial engineer is overseeing a production line upgrade, tasks such as installing new conveyors and training operators may have dependencies that require careful sequencing to prevent bottlenecks. By mapping out each task's timeline and dependencies, the project manager ensures that resources are available when needed, helping prevent downtime and resource conflicts.

Budgeting and resource allocation are finalized in the planning phase. Project costs are estimated, and budgets are created to cover equipment, labor, and material expenses. Resource allocation ensures that skilled personnel and necessary materials are assigned to each task at the right time. For instance, a facility layout redesign might require engineers, electricians, and construction workers at different stages. Planning ensures each team member is scheduled efficiently, reducing idle time and avoiding overspending.

Once planning is complete, the project moves to the **Execution phase**, where the actual work begins. In this phase, tasks are performed as per the project plan, and resources are utilized to create the deliverables outlined in the project's goals. For an industrial engineer managing a quality improvement initiative, execution might involve installing new quality control systems, training employees, and running pilot tests. Execution requires close coordination among team members and vendors to ensure tasks are completed on time and within budget.

Effective **communication** during execution keeps stakeholders informed about progress and any emerging issues. Regular meetings, status updates, and reports help project managers address issues in real time. For example, if installation delays occur due to supplier issues, the project manager can inform stakeholders, adjust the schedule, or arrange alternative suppliers. Clear communication prevents misunderstandings and ensures all stakeholders are aligned with the project's goals and timelines.

The **Monitoring and Controlling phase** runs concurrently with execution, tracking project performance against the plan. This phase involves assessing progress, quality, and spending to identify any deviations from the schedule, budget, or quality standards. Tools like Earned Value Management (EVM) allow project managers to measure actual performance against planned objectives, providing a quantitative assessment of project health. For instance, if a production line upgrade is running over budget, EVM would highlight this variance, prompting the project manager to review expenses and make adjustments to keep the project on track. Monitoring ensures the project remains aligned with its objectives and that any issues are corrected promptly.

In addition to monitoring costs, **quality control** ensures deliverables meet required standards. Inspections, tests, and audits verify that each project component functions as intended and that any defects are addressed immediately. For example, in a machinery installation project, testing the equipment's performance before full integration ensures it operates reliably and meets quality expectations. Quality control not only prevents rework but also enhances overall project efficiency by catching potential issues early.

The final phase, **Closing**, involves completing any remaining tasks, finalizing paperwork, and formally handing over deliverables. In this phase, the project manager conducts a final review to ensure all objectives have been met and that all tasks are complete. For example, in a facility upgrade project, closing would include a walkthrough to verify all installations, finalize contractor payments, and obtain sign-offs from stakeholders. This phase also includes a post-project evaluation, where lessons learned are documented. Reviewing what went well and what could improve helps industrial engineers refine their approaches for future projects.

The project life cycle provides industrial engineers with a clear roadmap, guiding them through each phase of a project from initiation to closing. By structuring projects in phases, engineers can manage resources efficiently, mitigate risks, and deliver successful outcomes.

Planning and Scheduling with Gantt Charts and CPM/PERT

Planning and scheduling are essential in project management, allowing industrial engineers to map out tasks, allocate resources, and track project timelines. **Gantt charts** and **CPM/PERT** (Critical Path Method/Program Evaluation and Review Technique) provide visual representations of project tasks, timelines, and dependencies, making complex projects easier to manage and monitor.

A **Gantt chart** is a horizontal bar chart that represents a project schedule over time. Each bar represents a specific task, with its length showing the task's duration and position indicating its start and end dates. For example, in a factory relocation project, tasks like site preparation, equipment setup, and quality testing would each have their own bar on the chart, visually showing when each task begins and ends. The Gantt chart's simplicity allows project managers to quickly see the project's overall timeline and identify overlapping tasks, helping manage tasks that can run simultaneously and reduce total project time.

Dependencies between tasks are essential to scheduling, and Gantt charts allow engineers to map these dependencies visually. For example, in a production line installation, a dependency might exist between completing electrical wiring and starting machine testing. Gantt charts show these dependencies with connecting arrows, making it clear which tasks must be completed before others can begin. This visual helps project managers identify potential bottlenecks and plan around dependencies to keep the project on schedule.

Critical Path Method (CPM) helps identify the longest path through a project, determining the minimum time required to complete it. CPM involves listing all tasks, determining durations, and identifying dependencies to map out possible paths. The longest path is the critical path, and any delay in tasks on this path delays the entire project. For instance, in a facility expansion project, tasks like obtaining permits, foundation work, and structural assembly may form the critical path. Monitoring these tasks closely ensures that delays are minimized, keeping the project within its planned duration.

PERT (Program Evaluation and Review Technique) is used when task durations are uncertain, providing three time estimates: optimistic, pessimistic, and most likely. This approach is particularly useful in projects with many variables, such as new product development. For example, an industrial engineer working on a product launch can use PERT to estimate time frames for product testing, certification, and production ramp-up. By averaging these estimates, PERT provides a more realistic timeline, accounting for uncertainties and reducing the risk of schedule overruns.

Combining Gantt charts with CPM/PERT gives project managers a comprehensive view of both task durations and critical dependencies. While Gantt charts offer a visual layout of the project schedule, CPM/PERT provides insights into timing risks and critical tasks. For instance, in a production line redesign, Gantt charts can show task overlap, while CPM highlights which tasks require close attention to prevent project delays.

With regular updates, Gantt charts, CPM, and PERT enable project managers to track progress, adjust schedules, and communicate effectively with stakeholders. These tools support proactive decision-making, allowing industrial engineers to allocate resources efficiently and adjust plans in response to unexpected delays.

Resource Allocation and Budgeting

Resource allocation and budgeting are crucial components of project management, ensuring that personnel, equipment, and financial resources are efficiently distributed to meet project objectives. In industrial engineering, effective resource management helps avoid delays, minimize costs, and optimize project outcomes.

Resource allocation involves assigning available resources to project tasks based on their requirements and timelines. Engineers begin by identifying the resources needed, including skilled personnel, machinery, materials, and tools. For example, in an equipment installation project, resources might include specialized technicians, lifting equipment, and safety gear. Once resources are identified, project managers allocate them to specific tasks, ensuring they are available when needed. Overlapping tasks with shared resources are scheduled to avoid conflicts, such as ensuring that a skilled technician is not required in two places at once.

Resource leveling is a technique used to balance resource availability with project requirements. When demand for resources fluctuates, leveling redistributes resources to minimize peak workloads and prevent underutilization. For instance, in a factory renovation, if multiple tasks require electricians, resource leveling would stagger these tasks to avoid overwhelming the available personnel. This approach helps keep workloads manageable, ensuring that resources are used efficiently across the project.

Budgeting establishes a financial framework for the project, outlining the funds available for each phase and task. Engineers estimate costs for labor, materials, equipment, and any subcontracted work. In a manufacturing facility upgrade, budgeting might include costs for machinery procurement, labor, and installation materials. Engineers break down the total project cost into smaller, manageable budgets assigned to individual tasks or phases, providing a clear financial roadmap for the project.

Cost estimation is central to accurate budgeting. Project managers assess potential expenses by analyzing past projects, obtaining quotes, or using standard industry rates. For example, if a project requires renting a crane, the cost estimate would include rental fees, operator costs, and fuel expenses. Detailed cost estimation prevents unexpected expenses and ensures that the budget aligns with project scope.

To monitor spending, project managers use **cost control measures**, tracking expenses against the budgeted amount. For instance, if a project requires additional materials due to design changes, cost control measures ensure that these expenses are documented and do not exceed the contingency budget. Cost tracking software and regular budget reviews help keep finances in check, ensuring that the project remains within budget constraints.

Contingency budgeting provides a safety net for unexpected costs, allowing flexibility without compromising project scope. Project managers typically set aside a percentage of the total budget as a contingency fund. For instance, in an industrial construction project, contingencies might cover unforeseen costs like weather delays or unexpected site conditions. Contingency funds help absorb financial shocks, reducing the risk of budget overruns.

Effective resource allocation and budgeting prevent resource shortages and financial strain. By assigning resources strategically and monitoring costs, industrial engineers ensure projects remain on schedule and within budget, maximizing efficiency and reducing risks.

Risk Assessment and Mitigation

Risk assessment and mitigation are essential in project management, allowing industrial engineers to identify, analyze, and address potential issues that could impact project outcomes. By anticipating risks early, project managers develop strategies to minimize their impact, ensuring that the project proceeds smoothly and efficiently.

Risk identification is the first step, involving a thorough examination of project activities to pinpoint potential risks. Engineers consider internal risks, such as equipment breakdowns, skill gaps, or schedule conflicts, and external risks, like supply chain delays, regulatory changes, or weather disruptions. For example, in a facility expansion project, risks might include delays in material deliveries or unexpected site conditions. By identifying risks at the outset, project managers prepare for challenges before they arise.

Once risks are identified, **risk analysis** assesses the likelihood and potential impact of each risk. Engineers use qualitative methods, like risk matrices, to categorize risks as high, medium, or low based on probability and severity. For instance, a risk matrix may classify the probability of labor shortages as medium but assign a high impact if critical skills are required. Quantitative methods, like Monte Carlo simulations, estimate potential delays or cost overruns, providing a statistical basis for risk management. This analysis allows project managers to prioritize high-impact risks, focusing efforts where they are most needed.

Risk mitigation strategies are developed for each identified risk, outlining specific actions to reduce likelihood or lessen impact. Mitigation strategies can involve preventive measures, contingency plans, or adaptive solutions. For example, in a project with tight deadlines, risk mitigation might include securing backup suppliers to prevent delays in material delivery. Preventive measures help avoid risk occurrence, while contingency plans prepare the team to respond effectively if a risk materializes.

Contingency planning is crucial for managing unforeseen events. Engineers allocate resources, budgets, and backup plans to address issues like equipment failures, labor shortages, or design changes. For instance, in a machinery installation project, contingency planning might involve budgeting for potential overtime if installation delays occur. These plans ensure the project can adapt to unexpected challenges, minimizing disruptions and keeping the timeline intact.

Monitoring and control are ongoing, allowing project managers to track risks and respond proactively. Engineers regularly review project status, evaluate risk indicators, and adjust mitigation strategies as needed. For example, if a supply chain delay appears likely due to external factors, the project manager might expedite orders or adjust the project schedule. Continuous monitoring ensures that risks are managed effectively, preventing small issues from escalating into major disruptions.

Effective risk assessment and mitigation allow industrial engineers to manage uncertainty, protect budgets, and maintain schedules. By anticipating risks, analyzing impacts, and implementing preventive measures, engineers create resilient projects that withstand challenges and achieve successful outcomes.

CHAPTER 11: INVENTORY MANAGEMENT AND CONTROL

Types of Inventory and Inventory Costs

In inventory management, understanding different types of inventory and their associated costs is essential for optimizing storage and ensuring a smooth production process. Each type of inventory serves a specific purpose within the supply chain, from raw materials to finished goods, and each incurs costs that impact a company's overall profitability and efficiency.

Raw materials are the basic inputs used to produce a finished product. For a furniture manufacturer, raw materials might include wood, metal fasteners, and varnish. Raw materials are typically sourced from suppliers and stored until they're needed in production. Managing raw material inventory carefully is critical because shortages can halt production, while excess can increase holding costs. Engineers often use reorder point systems to maintain optimal raw material levels, ensuring production continuity without overstocking.

Work-in-process (WIP) inventory includes items that have entered the production process but are not yet complete. For example, in an automotive plant, partially assembled vehicles on the assembly line are part of the WIP inventory. WIP inventory ties up resources, including labor and machinery, and reflects production efficiency. High WIP levels can indicate bottlenecks, while low levels suggest streamlined production. Managing WIP effectively involves balancing production speed with resource allocation to prevent bottlenecks and optimize throughput.

Finished goods inventory consists of products that are ready for sale to customers. This type of inventory acts as a buffer between production and customer demand, allowing companies to respond quickly to orders. For example, in a consumer electronics warehouse, ready-to-ship items like smartphones and laptops are part of the finished goods inventory. Balancing finished goods inventory involves meeting customer demand without overproducing, as excess stock increases storage costs and risk of obsolescence, especially in rapidly changing markets.

Maintenance, repair, and operations (MRO) inventory refers to items used in production and facility maintenance but not directly involved in creating the final product. MRO inventory includes spare parts, cleaning supplies, and safety equipment. For instance, a manufacturing plant might keep a supply of machine lubricants and replacement belts to ensure smooth operations. Proper MRO inventory management prevents production downtime due to equipment failure, allowing for timely repairs and maintenance.

Each inventory type incurs **inventory costs**, which are classified into several categories. **Holding costs** are expenses related to storing and maintaining inventory over time. These include warehousing costs, insurance, and costs for maintaining a controlled environment if required. For example, a warehouse storing perishable goods incurs higher holding costs due to refrigeration needs. Holding costs increase with inventory levels, encouraging companies to minimize excess stock to reduce expenses.

Ordering costs are the expenses associated with replenishing inventory. This includes administrative costs, transportation, and the time spent placing and receiving orders. For a retail store, ordering costs cover activities like placing purchase orders, receiving shipments, and processing invoices. Smaller order quantities can reduce holding costs but increase ordering frequency, raising ordering costs. Inventory managers often use the Economic Order Quantity (EOQ) model to find the optimal order size that minimizes the combined ordering and holding costs.

Stockout costs occur when a company runs out of a needed item, leading to lost sales, delayed production, or strained customer relationships. For example, if a car manufacturer runs out of a key component, production halts, causing delays and potentially damaging the company's reputation. Stockout costs are both direct (lost revenue) and indirect (customer dissatisfaction). To avoid stockouts, companies may carry safety stock or implement just-in-time inventory practices, balancing availability with the costs of holding additional stock.

Setup costs are specific to production settings and arise when switching from producing one item to another. For example, in a printing facility, setup costs involve adjusting equipment and cleaning presses between different print jobs. Setup costs are higher in environments that frequently switch between products. By minimizing setup time, often through techniques like Single-Minute Exchange of Dies (SMED), companies reduce the frequency and expense of setups, leading to more efficient production and lower overall inventory costs.

Obsolescence costs occur when inventory items lose value over time due to changing demand or technological advances. For example, in the electronics industry, holding onto outdated models increases the risk of obsolescence as customer preferences shift toward newer products. Obsolete inventory results in sunk costs, as unsold items often require discounting or disposal. Effective forecasting and inventory turnover strategies help companies reduce obsolescence risk, ensuring inventory aligns with current demand.

Shrinkage costs represent losses due to theft, damage, or administrative errors. Shrinkage is common in retail and warehouse environments, where items can go missing or be damaged during handling. Shrinkage reduces inventory value and is often mitigated through security measures, regular audits, and accurate tracking systems. Industrial engineers implement robust inventory management systems and technologies like RFID to minimize shrinkage and improve inventory accuracy.

Understanding inventory types and costs allows engineers to optimize inventory strategies, balancing availability with cost efficiency and ensuring that resources are used effectively within the supply chain.

Economic Order Quantity (EOQ) Models

The Economic Order Quantity model is a fundamental tool in inventory management, designed to determine the optimal order size that minimizes total inventory costs. EOQ helps balance ordering costs and holding costs, ensuring that inventory levels are sufficient to meet demand without incurring unnecessary expenses. In many cases, EOQ is used for items with relatively stable demand and known lead times, making it particularly effective for managing stock levels in manufacturing, retail, and supply chain contexts.

The **EOQ formula** is derived from the trade-off between ordering costs and holding costs. The EOQ formula is typically expressed as EOQ = $\sqrt{(2DS)/H}$, where D represents annual demand, S is the ordering cost per order, and H is the holding cost per unit per year. This formula calculates the ideal order quantity that minimizes total inventory costs by finding the point where the cost of ordering equals the cost of holding inventory. For instance, if a company produces a product with predictable demand, EOQ can guide order size to avoid frequent reordering and excessive stockpiling, balancing inventory needs with cost efficiency.

Ordering costs are incurred each time an order is placed, covering expenses like order processing, shipping, and receiving. If a company places frequent, small orders to avoid holding excess inventory, these ordering costs add up quickly, making small orders inefficient. Conversely, **holding costs** represent the expenses of storing inventory, including warehousing, insurance, and the risk of obsolescence. Large orders reduce ordering frequency but increase holding costs due to the additional space and handling required. EOQ seeks the balance between these costs, achieving a cost-effective middle ground.

In practice, EOQ models assume **constant demand** and fixed ordering and holding costs. For example, in a warehouse that stocks fast-moving consumer goods with steady demand, EOQ helps determine the ideal order size that minimizes both warehouse space and replenishment efforts. However, if demand fluctuates or lead times vary, EOQ may need adjustments to account for seasonal demand or supply chain variability. Some companies also incorporate safety stock calculations into EOQ models to prevent stockouts, especially for items with critical availability requirements.

One variation of the EOQ model is the **Economic Production Quantity (EPQ)** model, which applies when a company produces items internally rather than ordering them. EPQ considers production rates and includes the cost of setting up and shutting down equipment for each production run. For instance, in a metal fabrication plant, EPQ helps determine the optimal production run size for specific components, minimizing downtime and setup costs. This model is useful for manufacturers that need to balance production efficiency with inventory control, ensuring they meet demand without overproducing.

EOQ models also support **quantity discount analysis**, where suppliers offer price reductions for larger orders. When discounts are available, companies evaluate the potential savings from larger order sizes against the increased holding costs. For instance, if a supplier offers a 5% discount for orders above a certain quantity, the company calculates whether the reduced unit price offsets the higher holding cost associated with larger inventory levels. This analysis ensures companies take advantage of cost-saving opportunities without overspending on unnecessary stock.

EOQ models help streamline **inventory replenishment** processes by establishing a routine for ordering. Many companies set reorder points based on EOQ, automatically triggering orders when inventory reaches a specific level. For example, in a retail setting, when stock drops to a predefined reorder point, EOQ-based systems generate purchase orders, keeping shelves stocked and minimizing manual intervention. Automated EOQ-based ordering simplifies inventory management and maintains consistent stock levels, reducing the risk of stockouts.

Despite its simplicity, EOQ is an effective tool for maintaining inventory efficiency, especially when combined with modern software and data tracking systems. Inventory management software integrates EOQ calculations with real-time inventory data, allowing managers to monitor stock levels and automatically adjust order sizes. In industries with stable demand patterns, EOQ provides a reliable framework for balancing ordering and holding costs, enabling companies to optimize resources and minimize total inventory expenses.

Just-in-Time (JIT) and ABC Analysis

Just-in-Time and ABC Analysis are two inventory management techniques that enhance efficiency by optimizing stock levels and prioritizing inventory based on importance. Both methods aim to reduce excess inventory and improve resource allocation, supporting lean and cost-effective operations.

Just-in-Time is a lean inventory strategy focused on minimizing inventory levels by ordering items only when they are needed for production or sale. JIT requires precise scheduling and close coordination with suppliers to ensure materials arrive just in time to meet demand. In a manufacturing environment, for instance, raw materials and components are delivered directly to the assembly line, reducing storage needs and minimizing holding costs. This approach reduces waste and improves cash flow, as companies avoid tying up funds in surplus inventory. However, JIT requires a highly reliable supply chain, as any delay can halt production.

A key element of JIT is **demand forecasting**, as accurate predictions of customer needs are essential to avoid stockouts. If a company consistently underestimates demand, it risks running out of stock and disappointing customers. Conversely, overestimating demand leads to unnecessary inventory accumulation, defeating the purpose of JIT. For example, an electronics manufacturer using JIT monitors historical sales data, market trends, and real-time demand signals to adjust order quantities and ensure a smooth production flow. By aligning stock levels closely with demand, JIT reduces waste and improves efficiency.

Supplier partnerships are critical in JIT to maintain a responsive and flexible supply chain. Companies often work closely with suppliers to ensure timely deliveries and high-quality materials. For example, in the automotive industry, suppliers may deliver parts several times a day directly to the production line. Establishing long-term, trust-based relationships with suppliers reduces the risk of delays and enables JIT to operate smoothly. Some companies implement **vendor-managed inventory (VMI)**, where suppliers manage their products within the customer's facility, taking responsibility for stock levels and reordering as needed.

While JIT reduces holding costs, it requires thorough **risk management** to handle potential supply chain disruptions. Natural disasters, transportation delays, or quality issues can disrupt the timely arrival of goods. Companies using JIT often implement contingency plans, such as having backup suppliers or maintaining a minimal level of safety stock for critical items. For instance, a pharmaceutical company practicing JIT may keep emergency supplies of essential ingredients to avoid production stoppages. JIT's reliance on efficiency and reliability makes it ideal for industries with consistent demand and a stable supply chain.

ABC Analysis is another inventory management technique that classifies items into three categories—A, B, and C—based on their importance to the business. **Category A** items are high-value or high-demand products that represent a large portion of sales or revenue, typically making up around 20% of items but contributing to about 80% of value. These items require close monitoring and frequent reordering to prevent stockouts. In retail, for instance, a store might classify popular, high-margin products as Category A, ensuring that they are always in stock.

Category B items have moderate demand or value, requiring regular but not intense oversight. These items are important but less critical than A items. For example, in a hospital, commonly used but less expensive medical supplies, like bandages, might fall into Category B. Regular replenishment is necessary, but not as frequently as for Category A items.

Category C items are low-value products that make up a large portion of inventory but contribute minimally to sales. These items may require minimal monitoring and larger reorder quantities, as stockouts do not significantly impact operations. Office supplies in a corporate setting often fall into Category C, where large quantities are ordered infrequently to save on ordering costs.

ABC Analysis helps companies allocate resources based on the relative importance of inventory items. By focusing on A items, companies ensure that high-value or high-demand products receive priority in ordering, warehousing, and monitoring. This prioritization reduces the likelihood of stockouts for essential items while minimizing excess stock for less critical items. For example, a manufacturing plant might dedicate more storage space and management attention to Category A parts, while Category C items are kept in bulk storage.

Combining JIT with ABC Analysis enhances inventory control, as companies manage high-priority items with precision while maintaining flexibility for less critical stock. This approach optimizes inventory levels across all categories, improving efficiency and supporting lean operations.

Inventory Tracking and Management Systems

Inventory tracking and management systems are essential tools in inventory control, providing real-time visibility into stock levels, locations, and movement within the supply chain. These systems streamline operations by automating tasks such as receiving, picking, and restocking, reducing errors and improving efficiency. Modern inventory management systems often integrate with technologies like barcode scanning, RFID, and ERP systems, creating a comprehensive solution for managing inventory across various stages.

Barcode scanning is a fundamental technology in inventory tracking, allowing workers to scan items at each stage, from receiving to shipping. Each item or SKU (Stock Keeping Unit) is assigned a unique barcode, which the system reads to record its location and status in real time. For instance, in a distribution center, incoming shipments are scanned upon arrival, updating the system instantly with their location within the warehouse. As items move through picking and packing, barcode scans track each movement, ensuring that stock levels reflect current availability and reducing the risk of misplaced inventory.

RFID (Radio Frequency Identification) tags go beyond traditional barcodes, enabling automated, hands-free tracking of inventory. RFID tags contain a chip with product data that RFID readers can detect without direct line-of-sight. This allows large batches of items to be scanned simultaneously, speeding up inventory counts and reducing labor. In retail, for example, RFID systems can scan an entire pallet of products at once, updating stock records instantly and saving significant time compared to manual counts. RFID's real-time tracking capability enhances inventory accuracy, particularly for high-value or high-volume items, by providing continuous visibility into item locations.

Warehouse Management Systems (WMS) are software platforms that centralize inventory data and manage warehouse operations. WMS tracks stock levels, oversees order fulfillment, and optimizes storage layouts. For instance, a WMS can determine the most efficient placement of items within a warehouse based on demand frequency, minimizing travel time for pickers. WMS also integrates with barcode and RFID technologies, enabling workers to scan items and update stock levels instantly. The system provides alerts when items reach reorder points, ensuring that managers can initiate replenishment before stockouts occur. With a WMS, companies can maintain accurate stock records, improve warehouse productivity, and reduce storage costs.

Enterprise Resource Planning (ERP) systems extend inventory management beyond the warehouse, integrating inventory data with broader business functions like sales, procurement, and finance. ERP systems provide a unified view of inventory across multiple locations, helping companies manage stock levels more effectively. For example, a manufacturer with several warehouses and production sites can use an ERP system to monitor inventory levels at each location, reallocating stock as needed to meet demand. ERP systems also support demand forecasting, allowing businesses to anticipate inventory needs based on historical sales data and market trends, ensuring that inventory aligns with customer demand.

Cloud-based inventory management systems enable remote access to inventory data, allowing managers to monitor stock levels and make decisions from any location. Cloud solutions are especially valuable for businesses with distributed operations, as they allow for real-time visibility into inventory across multiple warehouses or retail sites. For example, a retailer with stores across different regions can monitor stock levels at each location remotely, reallocating products between stores to balance inventory and meet regional demand. Cloud-based systems also support real-time collaboration with suppliers, enabling seamless ordering and replenishment.

Automated inventory replenishment is another feature of advanced inventory management systems, using predefined parameters like reorder points and safety stock levels to trigger purchase orders automatically. For instance, if stock levels for a high-demand item reach the reorder point, the system can automatically generate a purchase order and send it to the supplier, ensuring continuous availability. Automated replenishment reduces manual intervention, speeds up restocking, and minimizes the risk of stockouts.

Inventory analytics within management systems provide insights into stock turnover rates, demand patterns, and seasonal trends. By analyzing this data, companies can optimize stock levels, adjust reorder points, and streamline procurement. For example, if analytics reveal that a particular product experiences high demand during specific months, the system can adjust reorder points accordingly, ensuring sufficient stock for peak seasons. These insights help businesses avoid overstocking or understocking, improving overall efficiency and reducing carrying costs.

Safety Stock Calculations and Reorder Points

Safety stock calculations and reorder points are fundamental concepts in inventory control, designed to prevent stockouts and ensure a smooth supply chain. By establishing buffer stock levels and reorder triggers, companies can maintain adequate inventory to meet demand fluctuations and avoid disruptions in production or sales.

Safety stock is the extra inventory held to cover unexpected demand spikes or supply delays. This buffer stock acts as a cushion, protecting against stockouts when demand exceeds forecasts or suppliers face disruptions. For instance, a grocery store may keep additional inventory of high-demand products like dairy or bread to prevent shortages during sudden demand surges. Safety stock is especially crucial for products with variable demand or long lead times, where stockouts could lead to lost sales or production delays.

The **basic safety stock formula** calculates safety stock by considering demand variability and lead time variability: Safety Stock = $Z \times \sqrt{(\sigma^2_d \times L + D \times \sigma^2_L)}$. In this formula, Z is the service level factor, representing the desired probability of avoiding a stockout, σ_d is the demand standard deviation, L is the average lead time, D is the average demand, and σ_L is the lead time standard deviation. For instance, if a company aims for a 95% service level, it selects a Z value of 1.65 to achieve that target. Calculating safety stock based on demand and lead time variability ensures that companies hold sufficient inventory to meet their desired service levels.

Service level is a key factor in safety stock calculations, determining the probability that stock will meet customer demand during a lead time period. Higher service levels require more safety stock, balancing the cost of holding extra inventory with the need to avoid stockouts. For example, in the medical supply industry, where stockouts can disrupt healthcare operations, companies may aim for a high service level of 99%, holding more safety stock to avoid disruptions. Conversely, non-essential items may have lower service level targets, reducing safety stock requirements and minimizing carrying costs.

Reorder points (ROP) indicate the inventory level at which a new order should be placed to replenish stock before it reaches critically low levels. The basic formula for reorder points is ROP = (Average Demand × Lead Time) + Safety Stock. For instance, if a company's average daily demand for a component is 50 units and lead time is 7 days, the reorder point, without safety stock, would be 350 units. Adding safety stock provides a buffer, ensuring that stock does not run out even if demand spikes during the lead time.

Demand forecasting is essential for setting accurate reorder points, as it provides a basis for estimating average demand. By analyzing historical demand patterns, companies can predict future demand and adjust reorder points to reflect current trends. For example, a retailer might increase reorder points before peak shopping seasons, ensuring sufficient inventory to meet higher demand. By aligning reorder points with forecasted demand, companies prevent stockouts while minimizing excess inventory.

Lead time variability impacts both safety stock and reorder points, as unpredictable lead times increase the risk of stockouts. Companies factor in lead time variability by adjusting safety stock levels and reorder points to account for potential delays. For instance, if lead times fluctuate due to supplier reliability or shipping conditions, a higher safety stock level mitigates the risk. Accurate lead time tracking helps companies update safety stock and reorder points dynamically, maintaining a balance between availability and inventory cost.

Periodic and continuous review systems determine how frequently reorder points are evaluated. In a periodic review system, inventory levels are checked at set intervals, such as weekly or monthly, with reorder points adjusted based on current stock levels and demand forecasts. In contrast, a continuous review system monitors inventory levels in real time, triggering replenishment orders as soon as stock reaches the reorder point. For high-demand items, continuous review ensures a quick response to stock level changes, reducing the chance of stockouts.

CHAPTER 12: COST ANALYSIS AND FINANCIAL DECISION-MAKING

Fundamentals of Cost Analysis

Cost analysis is a method for evaluating the expenses associated with a project, process, or decision to determine its financial viability and identify opportunities for cost savings. In industrial engineering, cost analysis provides a structured approach to assess, compare, and control costs, supporting effective financial decision-making in manufacturing, supply chain, and production environments.

Fixed and variable costs are fundamental to cost analysis, as they help define the total cost structure. Fixed costs, such as rent, insurance, and equipment depreciation, remain constant regardless of production volume. For example, a factory's rent does not change whether it produces 1,000 units or 10,000 units per month. Variable costs, on the other hand, fluctuate with production levels and include expenses like raw materials, labor, and energy. In a bakery, ingredients and hourly wages vary directly with the quantity of bread produced. Understanding fixed and variable costs allows managers to estimate total costs based on production levels, providing insight into cost drivers and break-even points.

The **break-even analysis** calculates the production level at which total revenue equals total costs, meaning the company covers all expenses without profit or loss. The break-even point (BEP) is calculated as BEP = Fixed Costs / (Selling Price per Unit - Variable Cost per Unit). This formula shows the minimum production volume needed to avoid losses. For instance, if a product has a fixed cost of $10,000, a selling price of $50, and a variable cost of $30, the break-even quantity would be 500 units. Break-even analysis helps managers set realistic sales goals, make pricing decisions, and assess whether a project is financially feasible.

Direct and indirect costs also impact cost analysis. Direct costs are easily traced to a specific product or project and include expenses like raw materials and direct labor. For example, in a furniture factory, wood and fabric are direct costs for each chair produced. Indirect costs, like facility maintenance or administrative salaries, are harder to assign to a single product and are often spread across multiple projects or departments. By separating direct from indirect costs, companies can assess the true cost of producing each item, aiding in pricing and product-line decisions.

Overhead allocation distributes indirect costs across various products or departments, ensuring accurate cost assessments. Overhead might include utilities, equipment maintenance, and administrative salaries. Engineers use methods like activity-based costing (ABC) to allocate overhead more precisely, assigning indirect costs based on the resources each product or department actually consumes. For example, in a machine shop, overhead costs for electricity might be allocated based on machine hours used by each product line. Accurate overhead allocation prevents underestimating or overestimating costs, leading to more accurate financial assessments.

Cost variance analysis compares actual costs with budgeted costs, identifying deviations and their causes. Variances can be favorable, where actual costs are lower than expected, or unfavorable, where actual costs exceed the budget. For example, if budgeted material costs for a project are $5,000 but actual costs reach $6,500, the unfavorable variance signals an issue, such as supplier price increases or inefficient material usage. By analyzing cost variances, managers can address root causes, adjusting processes to control expenses and improve cost efficiency.

Opportunity costs represent the potential benefits missed when choosing one option over another. In project selection, understanding opportunity costs helps assess the value of different choices. For example, if a factory has limited capacity and must choose between producing Product A with a higher margin or Product B with greater demand, the opportunity cost of selecting one product is the forgone profit from the other. Opportunity cost analysis ensures that resources are allocated to projects with the highest return, maximizing overall profitability.

Sunk costs are past expenditures that cannot be recovered and should not influence future decisions. For instance, if a company invests in a software system that turns out to be ineffective, the cost of the software is a sunk cost. Rational decision-making focuses on future costs and benefits rather than sunk costs, which are irrelevant to current choices. Recognizing sunk costs prevents companies from investing additional resources in unprofitable ventures based on previous losses, helping to allocate funds more effectively.

Life cycle costing (LCC) evaluates the total cost of ownership for an asset or project over its lifespan. LCC considers all costs, from initial purchase and installation to maintenance, operation, and eventual disposal. For example, in selecting machinery, LCC analysis includes purchase cost, maintenance expenses, and energy usage. By understanding the full cost trajectory, companies can make more informed decisions, selecting options that minimize long-term expenses. Life cycle costing is particularly useful in capital-intensive industries, where upfront investments impact costs over many years.

Incremental costing examines the additional costs and benefits associated with a specific decision, such as producing an extra batch of products. Incremental cost analysis helps managers assess whether the additional revenue generated justifies the added expense. For instance, if producing 100 extra units costs $2,000 and generates $2,500 in revenue, the decision is financially sound. Incremental costing supports quick decision-making in dynamic production environments, helping companies adjust output based on market demand.

Cost-Benefit Analysis and ROI

Cost-benefit analysis (CBA) is a financial decision-making tool used to compare the costs and benefits of a project, helping determine whether the investment is worthwhile. By quantifying both tangible and intangible factors, CBA allows industrial engineers to make informed choices that align with financial objectives. This method assigns monetary values to all costs and benefits associated with a project, creating a clear picture of its potential return.

To begin with, **direct costs**—such as materials, labor, and equipment—are calculated. For instance, in the decision to install a new conveyor system in a warehouse, direct costs include the purchase price, installation expenses, and labor required. These costs are straightforward and form the initial investment baseline. **Indirect costs**, like increased utility expenses or facility maintenance, are also included, as they can significantly impact the total expenditure over time. By including both direct and indirect costs, CBA provides a realistic view of the project's financial requirements.

Benefits are then quantified, which can include increased production capacity, improved quality, reduced downtime, and enhanced employee productivity. For example, a new conveyor system might shorten order fulfillment times, enabling the warehouse to process more orders daily. The financial value of this improvement is calculated based on increased revenue or cost savings. If the new system reduces labor needs by 10%, those labor savings become a quantifiable benefit. In addition to tangible benefits, intangible benefits, such as improved worker satisfaction or reduced error rates, can be considered, although they are more challenging to quantify directly.

One commonly used metric in CBA is the **net present value (NPV)**, which calculates the difference between the present value of benefits and the present value of costs over a specific timeframe. NPV accounts for the time value of money, recognizing that a dollar today is worth more than a dollar in the future. For instance, if a manufacturing project has projected savings over five years, NPV discounts these savings to reflect their value in today's terms. A positive NPV indicates that benefits exceed costs, supporting a decision to proceed.

Another crucial metric is **return on investment (ROI)**, which measures the profitability of an investment relative to its cost. ROI is calculated by dividing net benefits by total costs, expressed as a percentage. For example, if a project costs $50,000 and generates $70,000 in benefits, the ROI would be 40%. ROI helps compare multiple projects, guiding managers toward investments with the highest financial return. An ROI above 100% signifies that the project's benefits double its costs, signaling a highly favorable investment.

Payback period is also commonly used in CBA to determine how long it will take for a project's benefits to cover its costs. A shorter payback period indicates a quicker return, which can be advantageous in fast-paced industries or uncertain markets. For example, if a facility upgrade project has a payback period of three years, managers know that they will recover the investment within that time, making it financially appealing in the short term.

CBA also includes a sensitivity analysis to assess how changes in key assumptions affect project viability. If demand fluctuates or labor costs increase, sensitivity analysis examines the impact on NPV or ROI. For example, in a supply chain project, sensitivity analysis might test how changes in transportation costs affect total benefits, allowing managers to prepare for potential cost increases.

Breakeven Analysis and Profit Planning

Breakeven analysis and profit planning are essential tools for understanding when a project or product will become profitable. Breakeven analysis determines the production level or sales volume at which total revenue equals total costs, eliminating losses. This insight helps managers set realistic goals and informs decisions about pricing, costs, and output levels.

The **breakeven point (BEP)** is calculated by dividing fixed costs by the difference between the selling price per unit and variable cost per unit: BEP = Fixed Costs / (Selling Price - Variable Cost). This formula provides the minimum output required to cover costs. For example, if a manufacturing project has fixed costs of $100,000, a selling price of $20 per unit, and variable costs of $10 per unit, the breakeven quantity is 10,000 units. At this level, the company makes no profit or loss. Knowing the breakeven point is crucial for planning production schedules and setting realistic sales targets.

In addition to volume-based BEP, **dollar-based breakeven analysis** helps determine the revenue needed to reach breakeven. This is calculated by dividing total fixed costs by the contribution margin ratio, which is the percentage of each sale that contributes to covering fixed costs. For instance, if a product has a 50% contribution margin, a company with $200,000 in fixed costs needs $400,000 in revenue to breakeven. This approach aids companies in setting revenue targets that align with their fixed cost structure.

Breakeven analysis also supports **pricing decisions**. If the breakeven point is high, it may indicate that pricing needs adjustment or that cost-saving measures are necessary. For instance, if breakeven analysis reveals that a product requires an unfeasibly high sales volume, managers might explore ways to reduce variable costs, such as sourcing cheaper materials. Adjusting the product's selling price or lowering costs changes the breakeven point, making profitability more achievable.

Profit planning builds on breakeven analysis to determine the level of sales required to achieve a desired profit. This is done by adding the target profit to fixed costs in the breakeven formula. For example, if a company wants a profit of $50,000 in addition to covering $100,000 in fixed costs, they add the desired profit to the fixed costs before dividing by the contribution margin. This adjusted calculation shows how much the company needs to sell to reach its profit goal.

Profit planning also helps assess the feasibility of long-term projects, estimating how changes in sales, production costs, or market conditions affect profitability. If the company forecasts increased demand, profit planning evaluates whether the current cost structure supports growth or requires expansion. For instance, in an industrial project, profit planning might reveal the need for equipment upgrades to increase production capacity and meet higher sales targets.

- Together, breakeven analysis and profit planning give companies a roadmap for financial success, aligning production and pricing strategies with profit goals.

Financial Metrics for Industrial Engineers

Financial metrics are vital tools for industrial engineers, as they provide insights into the cost-effectiveness and profitability of projects and processes. These metrics guide decision-making, helping engineers evaluate options based on financial performance and allocate resources efficiently.

Cost per unit is a basic metric that calculates the total cost to produce a single unit of product. This includes both fixed and variable costs, divided by total production volume. For instance, if a factory produces 5,000 units at a total cost of $50,000, the cost per unit is $10. This metric helps identify whether production processes are cost-effective, and comparing cost per unit across different facilities can reveal potential cost-saving opportunities.

Gross margin is another important metric, calculated as (Revenue - Cost of Goods Sold) / Revenue, expressed as a percentage. Gross margin shows the profitability of a product after accounting for production costs, providing a clear picture of how much revenue is left to cover other expenses. For example, if a product sells for $100 with a cost of $60, the gross margin is 40%. High gross margins indicate efficient production and pricing, while low margins signal the need for cost reductions or pricing adjustments.

Return on Assets (ROA) measures a company's ability to generate profit from its assets. Calculated as Net Income / Total Assets, ROA indicates how efficiently a company uses its resources. For instance, if a facility generates $1 million in net income with $5 million in assets, its ROA is 20%. This metric is particularly useful for assessing the productivity of capital-intensive assets, such as machinery or equipment, helping engineers justify investments or reallocations to improve returns.

Net Present Value (NPV) assesses the profitability of a project by calculating the difference between the present value of cash inflows and outflows over time. NPV considers the time value of money, making it ideal for long-term projects. For example, in evaluating a five-year expansion project, NPV accounts for the initial investment and projected cash flows, providing a clear view of the project's total value in today's terms. A positive NPV suggests a profitable project, guiding managers toward financially sound decisions.

Internal Rate of Return (IRR) calculates the discount rate at which the net present value of future cash flows equals zero. IRR is often used alongside NPV to assess the profitability of investments. For instance, if a project's IRR exceeds the company's required rate of return, it indicates that the project is financially viable. IRR helps engineers prioritize projects based on expected returns, ensuring that resources are allocated to the most profitable options.

Payback period measures the time required to recover the initial investment of a project. Shorter payback periods are typically preferred, as they allow the company to recover costs quickly and reduce risk exposure. For example, if a machine costs $50,000 and generates annual savings of $10,000, the payback period is five years. This metric provides a simple assessment of risk, particularly in uncertain markets or with short product life cycles.

These financial metrics equip industrial engineers with the tools to evaluate projects and processes objectively, ensuring efficient resource use and alignment with financial goals. By understanding costs, profitability, and returns, engineers can make data-driven decisions that enhance financial performance.

CHAPTER 13: SYSTEMS ENGINEERING AND INTEGRATION

Principles of Systems Thinking and Systems Engineering

Systems thinking and systems engineering are foundational concepts in industrial engineering, focused on understanding complex systems as interconnected wholes rather than isolated parts. Systems thinking encourages a holistic approach, considering how individual components interact and influence the overall system. Systems engineering applies this perspective to design, analyze, and integrate systems that function efficiently and effectively.

Systems thinking starts with viewing a system as a set of interrelated parts that collectively produce a specific outcome. Rather than focusing on individual elements, systems thinking examines how these elements interact. For example, in a manufacturing system, each machine, worker, and process impacts the production flow. A bottleneck in one area can slow down the entire line, affecting output. Systems thinking identifies such interdependencies, making it possible to pinpoint root causes of inefficiencies and design solutions that improve the entire process rather than just one part.

Feedback loops are central to systems thinking, providing insight into how systems self-regulate and adapt over time. Feedback can be reinforcing (positive feedback) or balancing (negative feedback). In an inventory system, for example, a reinforcing feedback loop might occur when high demand prompts additional production, leading to overstock if demand drops suddenly. Balancing feedback, like reorder points in inventory management, keeps stock within desired levels by triggering orders when stock falls below a threshold. Understanding feedback loops allows engineers to anticipate system behavior, prevent issues, and create stable processes.

Another core principle of systems thinking is **emergence**, which occurs when system components work together to produce characteristics not found in individual parts. In a production plant, the combination of skilled workers, efficient processes, and reliable equipment can lead to high productivity—an emergent property that wouldn't be achievable by any component alone. By recognizing emergent properties, engineers can design systems where the whole exceeds the sum of its parts, creating efficiencies that go beyond individual improvements.

Boundaries are also essential in systems thinking, as they define what is inside or outside the system. Defining clear boundaries helps determine which elements influence the system directly and which factors are external. For instance, a supply chain system might include suppliers, warehouses, and distribution centers, but external factors like market demand and fuel prices exist outside the system boundaries. Identifying these boundaries helps engineers focus on the factors they can control while monitoring external factors that impact the system indirectly.

Systems engineering builds on systems thinking by using structured methods to design and manage complex systems throughout their lifecycle. It encompasses defining requirements, developing solutions, integrating components, and testing to ensure everything works together. Systems engineering begins with a clear definition of system requirements, capturing the needs of stakeholders. For instance, if an automotive manufacturer is developing a new production line, systems engineering starts with understanding production volume targets, quality standards, and safety requirements.

Functional decomposition is a method within systems engineering that breaks down complex processes into smaller, manageable functions. By dividing a system into its essential tasks, engineers can analyze and improve each function individually. For example, in an automated assembly line, functional decomposition might separate tasks like material handling, welding, and inspection. Each function is optimized separately, then integrated into the larger system, ensuring each task contributes effectively to the overall workflow.

Integration and interoperability are critical in systems engineering, ensuring that all parts of a system work together seamlessly. During the integration phase, each component—machines, software, sensors—is tested to verify compatibility and functionality. For instance, in a smart factory, machines with sensors must integrate with a central control system to enable real-time monitoring and automated adjustments. Ensuring interoperability means designing components that communicate and function cohesively, creating a system that operates as a unified whole.

Testing and validation ensure that systems meet performance and quality standards. Engineers conduct rigorous testing, first on individual components and then on the integrated system. In a new software system for managing production, engineers might test modules like inventory tracking, scheduling, and data reporting independently before verifying the entire system's functionality. This approach catches potential issues early, reducing the risk of failures once the system is live.

Life cycle thinking in systems engineering involves considering each phase of a system's life, from design and implementation to maintenance and eventual disposal. For a manufacturing system, life cycle thinking includes planning for future upgrades, assessing maintenance needs, and anticipating the end-of-life transition. Engineers may design systems with modular components that are easy to replace or upgrade, ensuring long-term sustainability. Life cycle thinking minimizes long-term costs and improves adaptability by accounting for future changes and maintenance needs.

Risk analysis is another essential component of systems engineering, as it identifies potential issues that could disrupt the system. Engineers assess risks associated with each component and process, implementing mitigation strategies where needed. For example, in a power plant, risk analysis might involve identifying points where system failure could lead to downtime, safety hazards, or financial losses. Mitigating these risks might include redundancy in critical components or establishing preventive maintenance schedules to reduce the likelihood of unexpected breakdowns.

Systems Integration in Complex Operations

Systems integration in complex operations involves linking multiple subsystems to function as a cohesive, unified whole. In large-scale industrial environments, integration ensures that machinery, software, and human resources work together seamlessly, enhancing operational efficiency and reducing downtime. Integration is especially valuable in sectors like manufacturing, logistics, and energy, where even minor disruptions in one part of the system can cascade through the entire operation.

One critical aspect of systems integration is ensuring **data compatibility** between different software systems. Many operations rely on various applications for tasks like inventory management, production scheduling, and quality control. Integrating these systems requires compatible data formats and communication protocols, enabling real-time data sharing across all components. For example, a manufacturing plant might integrate its ERP (Enterprise Resource Planning) system with a WMS (Warehouse Management System) to provide real-time updates on inventory levels. As materials move through the supply chain, data flows seamlessly between systems, improving visibility and reducing manual entry errors.

Communication protocols are foundational to integration, enabling different subsystems to "speak" to each other. Protocols like MQTT (Message Queuing Telemetry Transport) or OPC UA (Open Platform Communications Unified Architecture) facilitate reliable data exchange between industrial machines and control systems. In a complex assembly line, for instance, sensors on each machine transmit performance data to a centralized monitoring system through these protocols, allowing engineers to track machine health and efficiency in real time. By standardizing communication, protocols prevent data loss and ensure that every subsystem receives the information it needs to operate effectively.

Human-machine interfaces (HMIs) are another vital component in integrated operations, providing workers with intuitive controls to monitor and manage automated processes. HMIs act as a bridge between operators and machines, displaying key metrics like machine status, throughput, and error notifications. In a chemical processing plant, an HMI might show temperature, pressure, and flow rate across multiple reactors, giving operators the information they need to adjust parameters in real time. Well-designed HMIs improve the accuracy and speed of human decision-making, enhancing system efficiency and safety.

Systems integration also requires **hardware compatibility** to ensure physical components can connect and function as a single unit. In a logistics operation, for example, barcode scanners, conveyor belts, and automated storage systems need compatible wiring, control interfaces, and power sources to function together. Hardware integration eliminates bottlenecks, allowing items to move smoothly from receiving to storage and finally to dispatch. By ensuring hardware compatibility, engineers create an environment where each component contributes effectively to the overall workflow.

Testing and validation are essential to ensure that integrated systems meet performance standards. Engineers test each subsystem in isolation before conducting full-system integration tests. In a renewable energy plant, each turbine's performance is tested individually, followed by integration testing with the central grid. This process identifies potential issues early, minimizing the risk of costly failures once the system is live. Testing verifies that every subsystem performs as expected within the integrated setup, ensuring reliability and efficiency.

System maintenance in complex operations relies on integration to streamline monitoring and diagnostics. Integrated systems enable predictive maintenance, where data from sensors and control systems is analyzed to forecast potential failures. For instance, in an automotive manufacturing plant, sensor data from robotic arms indicates wear patterns, allowing engineers to schedule maintenance before failure occurs. Integration supports predictive maintenance by providing centralized access to performance data, reducing downtime and improving asset longevity.

Scalability is also a consideration in systems integration, especially for companies looking to expand. Scalable integration frameworks allow operations to grow by adding new machinery, production lines, or software modules without disrupting existing workflows. In a warehousing operation, adding automated picking robots should integrate with the existing WMS and conveyor systems without reconfiguring the entire system. Scalability ensures that as operational demands increase, the integrated system can expand accordingly without compromising performance.

Managing Interdependencies and Interfaces

Managing interdependencies and interfaces is essential in systems engineering, as it ensures that connected components work cohesively and that changes in one area don't disrupt the entire system. Interdependencies describe how subsystems rely on each other, while interfaces are the points of interaction where subsystems exchange information or physical goods. By carefully managing these aspects, engineers can prevent bottlenecks, reduce downtime, and optimize performance across complex operations.

Identifying interdependencies is the first step in managing them effectively. This involves mapping out which subsystems rely on each other for inputs, outputs, or data. For instance, in a pharmaceutical manufacturing plant, the mixing and packaging stages are interdependent. Any delay in mixing impacts packaging, which could lead to production backlogs. Identifying these interdependencies helps engineers understand how disruptions in one area affect other parts of the system, making it possible to create contingency plans that mitigate such risks.

Standardizing interfaces between subsystems is essential for efficient data and material flow. In a supply chain system, interfaces exist between inventory management, order processing, and logistics. Standardized interfaces allow seamless data exchange, ensuring that inventory levels automatically update when an order is processed. This prevents stock discrepancies and enables faster response times. Standardization often involves implementing protocols like REST APIs for data exchange, ensuring that information can flow between systems even if they're developed by different vendors.

Interface design focuses on how data or materials are transferred from one subsystem to another. Effective design considers factors like data accuracy, transfer speed, and user accessibility. For example, in an automated warehouse, the interface between robotic pickers and conveyor systems must transfer items efficiently without delay. Poor interface design can cause bottlenecks, as incompatible components struggle to transfer items at the desired rate. Engineers design interfaces to match the operational needs of both systems, ensuring compatibility and preventing delays.

Buffer zones help manage interdependencies by providing a buffer between subsystems with different processing speeds. In a production line, if the assembly stage operates faster than the inspection stage, a buffer area temporarily holds items until they can be inspected. Buffers reduce pressure on downstream processes, allowing each stage to operate at its optimal speed without causing delays. By managing interdependencies with buffers, engineers create a smoother workflow that minimizes the impact of variations in process times.

Monitoring and feedback systems are essential for tracking interdependencies in real-time. Sensors and monitoring tools provide live data on how each subsystem performs, identifying early signs of potential disruptions. For instance, in a power distribution network, monitoring voltage levels at various substations reveals dependencies between different areas. Any imbalance triggers immediate alerts, allowing engineers to take corrective action before the issue affects the entire grid. Real-time monitoring strengthens interdependency management by providing a continuous view of system performance.

Decoupling strategies reduce the impact of interdependencies, allowing subsystems to operate independently when necessary. Decoupling is achieved by designing subsystems that don't rely on constant input from others. For example, in a distribution network, warehouses may be decoupled from real-time production by maintaining adequate stock levels. Even if production temporarily halts, warehouses can continue to fulfill orders. Decoupling ensures stability, giving each subsystem flexibility to operate independently without disrupting the whole system.

Cyber-Physical Systems and Industry 4.0

Cyber-physical systems (CPS) and Industry 4.0 are transforming systems engineering, integrating digital technologies with physical processes to create intelligent, automated operations. In CPS, physical devices like sensors, actuators, and machines connect to digital control systems, enabling real-time monitoring, control, and feedback. Industry 4.0 expands on CPS by using advanced technologies like the Internet of Things (IoT), artificial intelligence, and big data analytics to optimize production, logistics, and maintenance.

In a **cyber-physical system**, physical components like sensors collect real-time data, which the digital system processes to make decisions. For example, in a smart factory, sensors on machinery monitor parameters like temperature, vibration, and speed. The data flows to a central control system that analyzes patterns, identifying signs of wear or potential failure. When conditions deviate from set thresholds, the system can automatically adjust machine settings or alert maintenance teams. By combining physical and digital components, CPS enables responsive, adaptive operations that maintain high productivity and reduce downtime.

IoT connectivity is foundational to Industry 4.0, allowing machines, tools, and systems to communicate across a shared network. In an automated warehouse, IoT devices on storage shelves track inventory in real time, updating the central system with each item's location and status. This level of connectivity enables optimized storage, faster picking, and minimized stockouts. IoT allows for seamless data flow across the entire facility, giving engineers a comprehensive view of operations and allowing quick adjustments in response to demand changes.

Predictive maintenance uses data from CPS to anticipate equipment failures before they occur. Machine learning algorithms analyze data patterns, like temperature fluctuations or power usage, to identify early signs of wear. For example, in a chemical plant, predictive maintenance might use CPS data to detect small pressure changes in pipelines, signaling a need for maintenance. By predicting failures before they disrupt operations, predictive maintenance reduces unplanned downtime and extends equipment life, saving time and costs.

Digital twins are virtual models of physical systems, created using data from CPS. They simulate real-world operations, enabling engineers to test changes and predict outcomes without disrupting actual processes. In automotive manufacturing, a digital twin of an assembly line can test new configurations or adjust machine speeds to maximize efficiency. By simulating adjustments virtually, digital twins help refine system performance, saving time and reducing risk.

Data analytics and AI are integral to Industry 4.0, transforming vast amounts of CPS data into actionable insights. AI algorithms analyze production data to optimize scheduling, identify bottlenecks and recommend improvements. For example, in a high-volume production facility, AI-driven data analytics can identify patterns in downtime events, suggesting adjustments to production schedules or maintenance routines that reduce delays. By leveraging data analytics, Industry 4.0 systems continuously refine processes, resulting in increased productivity and reduced waste.

Machine learning algorithms are also applied in **quality control**, where they detect anomalies and improve consistency. In a manufacturing line, machine learning systems can analyze images of products to identify defects, ensuring only high-quality items proceed through the system. This level of precision reduces rework and scrap, directly impacting cost savings and product reliability. Through continuous learning, these algorithms improve over time, refining their detection abilities and adapting to new quality standards.

Autonomous robotics and **automation** are key elements of Industry 4.0, particularly in applications that demand precision and speed. Cyber-physical systems use robotics to handle repetitive tasks, like assembly, packaging, or inspection, freeing human workers for higher-level decision-making. In a warehouse, autonomous mobile robots (AMRs) carry goods from storage to packing areas, following optimized paths based on real-time inventory data. Robotics reduce human error, speed up processes, and enhance flexibility, as they can be reprogrammed quickly to handle new tasks or adjust to demand shifts.

Finally, **security in cyber-physical systems** is critical, as increased connectivity opens new vulnerabilities. Engineers implement robust cybersecurity measures, including encryption, authentication protocols, and network segmentation, to protect data and maintain operational integrity. For example, in a power grid system, CPS components must be secured to prevent unauthorized access, which could disrupt service or compromise data integrity. Ensuring cybersecurity is essential for the safe and reliable operation of Industry 4.0 systems, as they rely on interconnected networks for optimal performance.

CHAPTER 14: DATA ANALYTICS AND INDUSTRIAL ENGINEERING

Basics of Data Collection and Analysis

Data collection and analysis form the backbone of effective decision-making in industrial engineering, allowing engineers to measure, monitor, and improve systems. Collecting relevant data and analyzing it accurately helps uncover patterns, predict outcomes, and optimize processes across manufacturing, logistics, and quality control.

Defining objectives is the first step in data collection. Engineers need clear goals for what they intend to measure and why. For example, if a factory aims to reduce equipment downtime, the objective could be to track machine performance data to identify causes of delays. Setting specific objectives helps in selecting the right data sources and collection methods, ensuring that the data gathered directly supports decision-making.

Once objectives are set, **identifying data sources** is essential. Industrial data typically comes from three main sources: operational data from machines and production systems, transaction data from business processes, and environmental data, like temperature or humidity, which can impact production. For example, in a quality control application, data might come from sensors on production lines, monitoring variables like temperature, pressure, and speed. Choosing appropriate data sources is crucial, as irrelevant data can lead to misleading analysis.

Data collection methods vary depending on the type of data needed. In many industrial environments, sensors and IoT (Internet of Things) devices collect data in real time, feeding information into centralized databases or data lakes. For example, a bottling plant might use sensors on filling machines to monitor speed and accuracy, sending data directly to a control system. Manual data collection methods, such as employee observations or checklists, may be used for non-automated tasks, like recording maintenance activities. Engineers select methods based on accuracy requirements, data volume, and available infrastructure.

Data quality is critical to reliable analysis, as errors or inconsistencies can distort findings. Engineers ensure data quality through validation checks, calibration, and standardized collection processes. For instance, in a chemical processing plant, calibrating sensors regularly ensures that temperature and pressure readings remain accurate. When collecting manual data, standardizing formats helps avoid entry errors and makes data easier to analyze. High-quality data leads to more trustworthy analysis, guiding better decision-making.

Preprocessing and cleaning the data is necessary before analysis. Preprocessing includes organizing raw data, handling missing values, and removing outliers that could skew results. For example, if a production line records unusually high readings during a one-time equipment failure, removing these outliers prevents them from affecting normal trend analysis. Engineers might also convert data into a consistent format or unit for easier analysis. This step ensures that data is standardized and accurate, providing a solid foundation for meaningful analysis.

Descriptive statistics are used to summarize data, offering an overview of current conditions. Key metrics like mean, median, and standard deviation provide insights into performance averages and variability. For example, calculating the mean cycle time for a production process shows the average time taken per unit, while standard deviation reveals how much variation exists. Descriptive statistics offer a quick snapshot, helping engineers understand baseline performance before diving into deeper analysis.

Data visualization is a important in data analysis, as it makes complex data easier to understand and interpret. Charts, graphs, and heatmaps highlight trends, patterns, and outliers, making it easier to spot issues or opportunities for improvement. For example, a time series chart showing daily production volumes might reveal seasonal patterns or recurring dips that suggest potential bottlenecks. Visualizations make data accessible to both engineers and decision-makers, facilitating clear communication and informed decision-making.

Trend analysis examines data over time, helping engineers identify patterns and predict future performance. For instance, tracking machine downtime over several months can reveal patterns that indicate maintenance needs. If downtime tends to spike every few weeks, it might suggest preventive maintenance intervals to minimize disruptions. Trend analysis helps engineers move from reactive to proactive management, reducing the likelihood of unexpected issues.

Root cause analysis uses data to identify the underlying causes of problems. Techniques like Pareto analysis or the fishbone diagram break down issues into contributing factors, highlighting areas that require attention. For example, if a production line experiences frequent defects, root cause analysis might reveal that a specific machine or step in the process consistently produces more errors. By pinpointing the root cause, engineers can implement targeted improvements, directly addressing the problem.

Predictive analytics take data analysis further by forecasting future events based on historical data. In predictive maintenance, machine learning models analyze past breakdowns and operating conditions to predict when a machine will need servicing. For instance, a model might learn that certain vibration patterns in a motor typically precede failures. Predictive analytics allows for timely interventions, reducing downtime and maintenance costs.

Statistical Methods for Process Improvement

Statistical methods are essential in industrial engineering for improving processes, identifying inefficiencies, and ensuring product quality. These methods provide a data-driven foundation to analyze variation, monitor performance, and implement changes that enhance productivity and reduce waste across production lines and operations.

Descriptive statistics serve as the first step in process improvement, summarizing data to provide insights into current conditions. Key metrics like mean, median, range, and standard deviation allow engineers to understand the average performance and variability within a process. For instance, tracking the average cycle time on an assembly line shows the typical time to produce one unit, while standard deviation highlights any inconsistencies. Descriptive statistics give engineers a clear baseline of how the process currently operates, making it easier to identify outliers or unusual patterns.

Process capability analysis evaluates how well a process meets specified tolerance limits or quality standards. Capability indices, like C_p and C_{pk}, measure the process's ability to produce items within defined limits. For example, in a machine shop where parts must meet strict dimensional tolerances, a high C_p value would indicate that the process has minimal variability relative to the specification range. C_{pk} considers both variability and process centering, revealing if the process consistently produces outputs near the target specification. By assessing process capability, engineers can determine if adjustments are needed to meet quality goals.

Control charts are key tools in Statistical Process Control (SPC), enabling continuous monitoring of process stability. Control charts plot data over time, with control limits representing the natural variation in the process. If data points remain within these limits, the process is considered stable. For example, in a beverage bottling line, control charts track fill levels across bottles. If fill levels start deviating from the control limits, engineers investigate to identify causes, like machine wear or calibration issues. Control charts help detect shifts or trends early, allowing for proactive adjustments that keep processes within desired limits.

Hypothesis testing helps engineers make data-based decisions when implementing process changes. By comparing sample data before and after an intervention, hypothesis tests determine if observed differences are statistically significant. For instance, if a production line introduces a new tool intended to reduce cycle time, engineers can use a t-test to compare average cycle times before and after implementation. If the results show a significant improvement, the change is deemed effective. Hypothesis testing provides confidence in process adjustments, ensuring changes are truly beneficial.

Design of Experiments (DOE) systematically tests multiple factors to understand their impact on process performance. DOE uses factorial experiments to evaluate the combined effects of variables on an outcome, allowing engineers to optimize processes efficiently. In a manufacturing setting, DOE might involve adjusting factors like temperature, pressure, and speed to determine their effect on product quality. By analyzing these interactions, engineers identify optimal settings that maximize performance while minimizing waste. DOE is particularly valuable for complex processes where multiple variables influence outcomes, as it reduces trial-and-error testing.

Pareto analysis is used to prioritize issues based on their frequency or impact. The Pareto principle, often called the 80/20 rule, suggests that roughly 80% of problems result from 20% of causes. In quality control, engineers use Pareto charts to identify the most common sources of defects or delays. For example, in an electronics assembly line, Pareto analysis might reveal that a small number of steps cause the majority of quality issues. By focusing on these key areas, engineers can achieve significant improvements with targeted efforts, maximizing the impact of their resources.

Regression analysis explores relationships between variables to understand how changes in one factor affect another. In industrial engineering, linear regression models are frequently used to predict outcomes like production output or defect rates based on input variables. For instance, regression might reveal how temperature variations impact material strength in a plastic molding process. By understanding these relationships, engineers can adjust conditions to achieve desired results, improving consistency and reducing variability.

Root cause analysis often uses statistical tools like fishbone diagrams or histograms to break down complex issues and pinpoint underlying causes. For example, if a plant experiences frequent machine breakdowns, engineers might use histograms to track the frequency of failures by machine or component type. This data helps identify patterns, leading to targeted maintenance or replacement programs that address root causes instead of treating symptoms.

Predictive Analytics in Industrial Engineering

Predictive analytics enables industrial engineers to forecast future events and optimize processes based on historical data. By using techniques like machine learning and statistical modeling, predictive analytics anticipates issues, identifies trends, and supports proactive decision-making in production, maintenance, and quality control.

Data preparation is the initial step in predictive analytics, as clean, organized data is essential for accurate forecasting. Engineers collect historical data from sources like sensors, machine logs, and quality inspection records, and preprocess it to remove inconsistencies, handle missing values, and normalize formats. For instance, in a manufacturing plant, engineers might prepare datasets showing machine runtimes, maintenance intervals, and defect rates. Proper data preparation ensures that the model uses relevant information, reducing the risk of misleading predictions.

Machine learning algorithms are at the heart of predictive analytics, using patterns in historical data to make accurate forecasts. Common algorithms include regression models, decision trees, and neural networks. In a supply chain setting, linear regression might predict demand based on historical sales and seasonal factors, while decision trees classify likely causes of machine failure based on usage patterns. Machine learning algorithms improve predictive accuracy by learning from large data sets, making them adaptable to complex, changing environments.

Predictive maintenance uses machine learning to forecast equipment failures before they occur. By analyzing data from sensors and historical breakdowns, predictive models can identify signs of wear or irregularities that indicate an impending failure. For example, in an automotive plant, sensors monitor vibration and temperature in assembly line motors. When the model detects unusual patterns, it alerts the maintenance team to schedule repairs, preventing unplanned downtime. Predictive maintenance reduces costly breakdowns and increases equipment lifespan, improving overall productivity.

Anomaly detection identifies unusual patterns or outliers in data, signaling potential issues in real time. In quality control, anomaly detection algorithms scan production data to catch deviations from normal behavior. For instance, in a pharmaceutical plant, sensors monitor mixing temperatures. If an anomaly detection algorithm finds temperatures outside acceptable ranges, it flags a potential quality issue. Engineers can intervene immediately, preventing defective batches from progressing further along the line. Anomaly detection minimizes waste and maintains product quality by spotting issues early.

Forecasting demand and inventory needs is another area where predictive analytics enhances efficiency. Demand forecasting models analyze sales trends, seasonality, and market data to predict future demand accurately. For example, in retail, a demand forecasting model might use past sales data, weather patterns, and holiday schedules to predict inventory needs. By aligning production schedules with demand forecasts, companies can prevent stockouts and reduce excess inventory, leading to better customer service and lower carrying costs.

Quality control benefits from predictive analytics by identifying variables that lead to defects. Machine learning models can analyze factors like raw material quality, operator settings, and environmental conditions to predict defect likelihood. In a semiconductor fabrication facility, predictive models monitor variables like temperature and humidity to forecast the probability of defects in silicon wafers. By adjusting these factors preemptively, engineers minimize defect rates and improve yield, leading to higher quality products.

Energy consumption forecasting is increasingly valuable as companies seek to reduce costs and environmental impact. Predictive models analyze historical energy usage data and operational conditions to anticipate future consumption patterns. In a factory with high energy demands, engineers can use these forecasts to optimize energy-intensive processes, such as scheduling production during off-peak hours. By managing energy use more effectively, companies can lower costs, improve sustainability, and maintain steady operational conditions.

Process optimization leverages predictive analytics to improve efficiency by suggesting adjustments based on historical performance data. For instance, in a food processing plant, predictive models might analyze production speed, material input rates, and defect data to recommend optimal production settings. By adjusting variables to match ideal conditions, engineers reduce waste and enhance productivity, continually refining the process for best results.

Real-time monitoring and predictive control combine to create an adaptive system that responds dynamically to changing conditions. In a smart manufacturing environment, predictive control models continuously analyze data from sensors, making adjustments in response to detected changes. For example, if a temperature increase in a chemical reactor is likely to affect output quality, predictive control can adjust input flows automatically to maintain optimal conditions. This approach enhances precision and responsiveness, maintaining quality while reducing manual intervention.

Data Visualization and Interpretation

Data visualization transforms complex datasets into clear, interpretable graphics, allowing industrial engineers to analyze performance metrics, detect patterns, and make informed decisions. In industrial environments, visualization tools simplify the understanding of large volumes of data, from production output to quality metrics, making it easier to spot trends, anomalies, and areas for improvement.

Charts and graphs like line charts, bar charts, and scatter plots are commonly used to visualize performance over time or compare variables. For instance, a line chart displaying daily production output allows engineers to spot seasonal trends or dips in productivity, highlighting areas that may need process adjustments. A scatter plot, on the other hand, can reveal relationships between variables, such as the correlation between machine speed and defect rate, guiding optimization efforts by illustrating clear patterns.

Heatmaps are particularly useful in identifying bottlenecks or high-error areas on production floors or within warehouses. In a distribution center, for example, a heatmap of picker movement shows which zones see the highest traffic, helping managers optimize storage layout to reduce travel times and improve efficiency. By visually highlighting problem areas, heatmaps provide a quick reference that guides layout adjustments or resource allocation.

Dashboards consolidate multiple data points into one interface, allowing real-time monitoring across operations. Dashboards are used in control rooms to display key performance indicators (KPIs) like cycle time, downtime, and defect rate. In a manufacturing facility, a dashboard may include live feeds from sensors, updating instantly to reflect the current operational status. Engineers can respond to trends or anomalies immediately, making dashboards critical for fast-paced decision-making in industrial settings.

Interpreting visualizations accurately requires understanding both the data source and the context. Engineers consider factors like time frames, external influences, and operational changes that might affect data trends. For instance, a sudden drop in production output may be related to maintenance activities rather than a systemic issue. Accurate interpretation of visual data prevents misguided conclusions and allows for effective, targeted interventions.

Histograms and control charts are essential for analyzing quality and process stability. Histograms show the frequency distribution of measurements, such as product weights, revealing variations or shifts. Control charts monitor whether processes remain within acceptable limits, flagging points where intervention is necessary. For example, in quality control, control charts reveal trends toward defects, allowing early adjustments to maintain product standards.

Applying Machine Learning Techniques for Process Insights

Machine learning (ML) techniques allow industrial engineers to analyze data patterns, forecast outcomes, and optimize processes by using algorithms that learn from historical data. In complex industrial environments, ML provides insights that traditional data analysis might miss, such as predicting equipment failures, optimizing production schedules, and reducing defect rates.

Supervised learning is a commonly used machine learning approach in industrial settings. With supervised learning, algorithms learn from labeled data to predict outcomes based on historical patterns. For example, in predictive maintenance, engineers train a supervised model with historical data on machine operating conditions and breakdown events. Once trained, the model can predict future failures based on real-time data, alerting maintenance teams to perform preventive maintenance before issues occur. This approach minimizes downtime and improves equipment lifespan.

Classification algorithms within supervised learning help sort data into predefined categories, making them useful for quality control applications. In a food processing plant, a classification algorithm might analyze images of products, identifying visual defects and sorting items into acceptable and reject categories. By automating defect detection, classification algorithms improve quality assurance while reducing manual inspection labor. Decision trees, support vector machines, and neural networks are commonly used for classification tasks, each with strengths depending on data complexity and volume.

Regression analysis is another supervised learning technique that predicts continuous outcomes, such as production output, based on influencing factors. In a manufacturing environment, engineers might use regression to estimate yield based on factors like temperature, pressure, and material quality. For example, in plastic molding, regression models predict how different temperature settings affect product strength, allowing operators to adjust conditions for optimal results. By quantifying relationships between variables, regression helps engineers refine processes, improving consistency and reducing waste.

Unsupervised learning finds hidden patterns within unlabeled data, which is particularly useful in large-scale industrial datasets. Clustering algorithms, a common unsupervised technique, group data based on similarities, revealing patterns in complex data. For instance, clustering in customer order data can help identify purchasing trends, guiding production planning and inventory stocking. In production, clustering might be applied to sensor data to identify abnormal operating conditions, alerting engineers to possible issues before they escalate.

Anomaly detection is another unsupervised technique used to identify unusual patterns that may indicate problems. By training a model on normal operational data, engineers can configure it to detect deviations that signal potential failures. In an assembly line, anomaly detection might monitor torque levels during screw fastening, alerting operators if readings deviate significantly from normal. This proactive approach reduces defect rates by catching issues early and allows engineers to correct problems without halting production.

Reinforcement learning (RL), an advanced ML technique, is used in industrial settings for process optimization and automation. RL algorithms learn through trial and error, adjusting actions based on feedback to maximize rewards. In a warehouse setting, RL might optimize the routes of autonomous robots, minimizing travel time while avoiding congestion. By continuously refining its actions, an RL model improves over time, adapting to dynamic conditions and optimizing performance. RL is particularly valuable in scenarios with complex, interdependent processes where manual optimization is challenging.

Natural Language Processing (NLP), another branch of machine learning, has applications in industrial engineering for analyzing unstructured data like maintenance logs, incident reports, and quality feedback. NLP algorithms process text data, extracting keywords, sentiment, or trends that reveal underlying issues. For instance, in a manufacturing plant, analyzing maintenance logs with NLP might uncover recurring issues with a specific machine type, prompting preventive measures. NLP allows engineers to leverage textual data effectively, gaining insights that support continuous improvement.

Feature engineering is a critical step in machine learning for industrial applications, as it involves selecting and transforming variables that the model uses to make predictions. For example, when building a predictive maintenance model, engineers might include features like machine age, usage hours, and vibration levels, each providing valuable context. Feature engineering ensures the model considers the most relevant data, improving its accuracy and reliability. By carefully selecting features, engineers make machine learning models more interpretable and effective.

Dimensionality reduction techniques, like Principal Component Analysis (PCA), simplify large datasets by reducing the number of variables while retaining essential information. In industrial datasets with hundreds of variables, dimensionality reduction helps streamline analysis without sacrificing accuracy. For instance, in quality control for a complex product, PCA might reduce dozens of measurements into a few principal components, making it easier to track overall quality trends. Dimensionality reduction supports efficient model training and improves interpretability.

Once models are trained and deployed, **model monitoring and maintenance** are essential to ensure continued accuracy. Data conditions often change over time, causing "model drift" that reduces effectiveness. In a production environment, changes in materials, machines, or workflows might alter data patterns, making the model less accurate. Regular monitoring and retraining keep models aligned with current conditions, ensuring consistent performance.

CHAPTER 15: SIMULATION AND MODELING IN INDUSTRIAL ENGINEERING

Types of Simulation Models and Applications

Simulation modeling in industrial engineering allows engineers to create virtual models of complex systems, testing different scenarios and optimizing processes without disrupting actual operations. By understanding and applying various types of simulation models, industrial engineers gain valuable insights into production systems, logistics, and facility layouts, helping improve efficiency, reduce costs, and minimize risks.

Discrete-event simulation (DES) is one of the most common types of simulation used in industrial settings, especially for processes with distinct, sequential events. DES models simulate each event in a process individually, such as the arrival of parts on a production line, assembly time, and inspection. Each event occurs at a specific point in time, allowing engineers to analyze the flow of materials and detect bottlenecks. For instance, in an automotive assembly line, DES tracks each vehicle as it moves through stages like welding, painting, and final inspection. By analyzing the time taken at each step, engineers can pinpoint bottlenecks or idle times, testing changes like adding a second inspection station to see if it improves throughput.

DES is also valuable in **logistics and warehousing**. In a distribution center, DES models simulate activities like order picking, packing, and dispatching, accounting for variables like worker availability and travel time. By simulating various layout configurations or staffing levels, engineers determine the optimal setup that maximizes efficiency and reduces order fulfillment time. This approach allows companies to fine-tune operations and adjust resource allocation based on demand fluctuations, leading to cost savings and faster delivery times.

Continuous simulation models, in contrast to DES, represent processes that change in a smooth, uninterrupted flow. These models are ideal for systems where variables shift constantly, such as chemical processing or thermal dynamics in manufacturing. In a refinery, for example, continuous simulation tracks how changes in temperature or pressure affect the final product's quality, allowing operators to adjust parameters in real time to ensure optimal conditions. Continuous simulations help in processes where precise control is necessary to maintain quality and consistency, as engineers can model various operational scenarios and preemptively adjust to maintain steady-state conditions.

Agent-based simulation (ABS) focuses on modeling individual agents—such as machines, people, or vehicles—that interact within a system. Each agent in ABS follows a set of rules, responding to others' actions and adjusting behavior accordingly. In a factory floor simulation, ABS might represent each worker, machine, and material handling robot as an agent. Engineers observe how these agents interact, testing changes like reallocating tasks or adjusting machine availability. ABS is particularly useful in understanding complex systems where individual actions influence overall performance, such as worker productivity or autonomous vehicle paths in a warehouse.

ABS is also applied in **supply chain management** to model interactions between suppliers, manufacturers, and distributors. For example, in a global supply chain, each agent represents a different supplier or facility, with rules for lead time, production capacity, and shipping constraints. By testing various demand levels, engineers can observe how the system adapts, identifying vulnerabilities such as overreliance on a single supplier. ABS provides insights into dynamic interactions within the supply chain, enabling companies to improve resilience and responsiveness.

System dynamics simulation models focus on feedback loops and accumulations within a system, offering a high-level view of how different parts of a system influence each other over time. System dynamics is often used in strategic decision-making, particularly where long-term consequences are crucial, such as inventory planning or workforce management. In inventory management, system dynamics models simulate the effects of varying order quantities, lead times, and demand fluctuations over an extended period. Engineers can test strategies like increasing safety stock or adjusting order frequency, understanding the cumulative impact of these decisions on overall inventory costs and stockout risks. System dynamics is ideal for strategic planning, offering insights into how changes ripple through a system over time.

Monte Carlo simulation uses random sampling to predict outcomes, particularly useful when uncertainty is high. By running thousands of simulations with different random inputs, Monte Carlo models produce a range of possible outcomes, showing probabilities for each scenario. For instance, in a project management context, Monte Carlo simulations assess the likelihood of meeting deadlines given various risk factors like delays in material supply or resource availability. By understanding these probabilities, managers make informed decisions about scheduling, resource allocation, and contingency planning, improving project reliability.

In **quality control and risk assessment**, Monte Carlo simulation estimates the likelihood of defects or failures, helping companies set realistic quality standards and minimize risks. For example, in an electronics manufacturing process, engineers might simulate the probability of component failure under different stress levels, identifying areas that need reinforcement or redesign. Monte Carlo analysis helps quantify risk in uncertain situations, providing valuable input for decisions that balance cost, quality, and reliability.

Hybrid simulation models combine different types of simulation, such as integrating DES and ABS or DES with Monte Carlo, to tackle complex systems that require multiple approaches. In a hospital setting, a hybrid model might combine DES for patient flow through emergency rooms with ABS to represent interactions between medical staff, patients, and equipment. By modeling both patient queues and individual agent behaviors, hybrid simulations capture the intricacies of healthcare environments, allowing hospital administrators to optimize staffing levels, patient flow, and equipment usage to improve overall service quality.

Virtual and augmented reality simulations are important for modeling physical spaces and workflows. These simulations create immersive environments where engineers and operators can interact with virtual replicas of their systems. For example, a virtual reality (VR) simulation of a new production line allows operators to "walk through" the setup, observing layout and flow before construction begins. VR simulations offer hands-on insights into potential issues like safety hazards or inefficient layouts, enabling engineers to refine designs before implementation, reducing costly modifications later.

Simulation models offer a versatile and detailed approach to analyzing, optimizing, and preparing for various industrial engineering challenges. By selecting the right model type—whether DES, continuous simulation, ABS, system dynamics, Monte Carlo, hybrid models, or virtual reality—engineers tailor their approach to specific operational needs, building systems that are efficient, resilient, and adaptable to real-world demands.

Discrete-Event Simulation and Process Modeling

Discrete-event simulation (DES) is a popular technique in industrial engineering that models complex processes by breaking them down into individual events that occur in sequence. Each event represents a specific action or change, such as the arrival of a part, the start of an assembly operation, or the completion of a quality inspection. DES allows engineers to analyze and optimize processes by tracking these events over time, capturing dependencies, identifying bottlenecks, and testing modifications before making changes to real-world systems.

In **manufacturing**, DES models the flow of materials and products through production lines, capturing each step in the process. For example, on an automotive assembly line, DES models each stage—welding, painting, quality inspection, and assembly—as separate events. Engineers can then test adjustments, such as changing staffing levels, altering equipment speed, or adding parallel workstations, to see how these changes impact overall throughput. DES highlights where delays occur, such as when one station completes work faster than the next, causing bottlenecks. By simulating various configurations, engineers optimize the flow and reduce idle time, improving productivity and reducing costs.

Warehousing and logistics also benefit from DES by modeling the movement of goods, equipment, and personnel. In a distribution center, DES can simulate events such as product arrival, picking, packing, and dispatching, tracking the time required for each task. For instance, if product demand spikes, DES helps determine whether the current setup can handle increased volume or if more staff and equipment are needed. Engineers can simulate changes in warehouse layout, introducing conveyor belts or automated picking robots, to identify the most efficient design. By experimenting with different configurations, DES ensures that operations meet demand while minimizing travel time and maximizing worker productivity.

In **service industries**, DES improves process efficiency by simulating customer interactions and service delivery times. For example, in a hospital emergency room, DES models patient arrivals, triage, treatment, and discharge as individual events, tracking patient wait times and treatment durations. If a hospital wants to add more beds or staff, DES simulates these changes, showing their impact on patient flow and wait times. By visualizing the effects of resource allocation, DES enables hospital administrators to optimize service levels, improve patient experience, and reduce operational costs.

Supply chain management uses DES to model the movement of goods from suppliers to end customers, accounting for production lead times, transportation delays, and inventory levels. For instance, a company might use DES to simulate different supplier scenarios, testing the impact of faster shipping or larger order quantities on inventory costs and service levels. DES helps companies understand the trade-offs between lead times and stock levels, allowing them to balance costs with customer service requirements. By optimizing each event in the supply chain, DES ensures that products move efficiently from one stage to the next, minimizing delays and maximizing responsiveness.

DES is also valuable for **resource allocation** and **scheduling** by modeling each job or task as an event and determining the optimal allocation of resources like labor, machinery, and time. In construction, for example, DES models tasks like site preparation, foundation work, and structural assembly, each with different resource requirements and durations. By analyzing how resources are utilized in each phase, DES highlights potential inefficiencies and suggests scheduling adjustments that make better use of available resources. This level of detail allows project managers to schedule resources dynamically, adapting to changing conditions and preventing resource shortages or downtime.

One of the greatest advantages of DES is its ability to test **"what-if" scenarios** without disrupting actual operations. Engineers can adjust variables such as task durations, resource availability, and equipment capacity to see how changes impact the system. For example, in a pharmaceutical plant, DES might simulate changes in batch production size, testing whether larger batches reduce downtime or cause bottlenecks in packaging. This experimentation provides a risk-free way to explore different process configurations, helping engineers make data-driven decisions that enhance efficiency.

DES also provides insights into **system performance metrics** like cycle time, lead time, and utilization rates, offering a detailed view of how each component contributes to overall performance. For instance, in a telecommunications call center, DES tracks metrics such as average call wait time, handling time, and agent utilization. These metrics guide decisions on staffing levels, call routing, and scheduling to improve customer service.

Through discrete-event simulation, industrial engineers gain a deeper understanding of process dynamics, allowing them to design and optimize complex systems that are both efficient and adaptable.

Analyzing Simulation Data for Decision Making

Analyzing simulation data provides industrial engineers with critical insights for decision-making by transforming raw model outputs into actionable information. Once a simulation model, such as a discrete-event or system dynamics simulation, is complete, the data generated includes details on performance metrics, system behavior, and potential improvements. Analyzing this data allows engineers to make informed decisions that enhance efficiency, reduce costs, and improve overall system performance.

Key performance indicators (KPIs) are central to simulation data analysis, as they offer measurable insights into the system's effectiveness. KPIs vary by application, but common metrics include cycle time, throughput, resource utilization, and lead time. In a production setting, for example, KPIs might track the average time it takes for a product to move from start to finish on an assembly line. By focusing on KPIs, engineers can identify areas where the process falls short, such as high cycle times or low utilization, and implement targeted improvements to address these inefficiencies.

Bottleneck analysis is often conducted using simulation data, identifying areas in a process that restrict flow and reduce throughput. In a factory producing electronics, simulation data may reveal that certain workstations have significantly higher processing times than others, causing delays in subsequent steps. By isolating these bottlenecks, engineers can test solutions such as increasing machine capacity or reallocating labor, directly addressing constraints. Bottleneck analysis is essential for optimizing flow and ensuring that each process stage operates as efficiently as possible.

Scenario comparison enables engineers to analyze different configurations and choose the most effective approach. For instance, in a warehouse setting, simulation data might compare scenarios where additional pickers are added versus scenarios where automation is increased. By running multiple simulations, engineers can compare outcomes across scenarios, evaluating metrics like order fulfillment time, labor costs, and storage utilization. Scenario comparison ensures that decision-makers have data-driven options to choose from, supporting strategic adjustments without the risk of implementing costly changes prematurely.

Statistical analysis of simulation data is used to confirm whether observed differences are significant. Hypothesis testing and confidence intervals allow engineers to determine if a change in the process, such as altering equipment speed or reordering task sequences, produces meaningful improvements. In a bottling plant, for example, statistical tests might verify whether faster conveyor speeds significantly reduce cycle time without increasing defects. Statistical analysis lends rigor to simulation results, ensuring that conclusions are based on solid evidence rather than random variation.

Sensitivity analysis examines how sensitive the simulation outcomes are to changes in key parameters, providing insights into system stability and risk. Engineers adjust variables like demand levels, processing times, or resource availability to observe how the system responds. In a logistics network, for example, sensitivity analysis might involve varying shipping lead times to see how delays affect overall delivery performance. By understanding which variables impact the system most, engineers can focus on managing these critical factors and build resilience against disruptions.

Visualization of simulation data aids in interpreting complex datasets, as graphs, charts, and heatmaps reveal trends, bottlenecks, and anomalies clearly. For instance, a line chart might illustrate cycle time trends over a simulated month, showing where and when delays occur. Heatmaps in a warehouse simulation might show areas of congestion or high picker traffic, indicating areas for layout improvement. Visualization makes simulation data accessible, providing a quick yet thorough overview that supports better decision-making.

Cost-benefit analysis uses simulation data to evaluate the financial implications of different decisions, balancing the costs of changes with expected benefits. For example, in a manufacturing plant, simulation data may show that adding an extra machine improves throughput, but at a significant cost. Cost-benefit analysis calculates the payback period and long-term savings, helping managers decide if the investment justifies the expense. This analysis ensures that resources are allocated effectively, maximizing financial returns while maintaining operational efficiency.

Simulation data also provides **forecasting insights**, predicting system behavior under various future conditions. In a hospital emergency department, simulation data might forecast patient wait times based on seasonal demand fluctuations, staffing levels, or bed availability. These forecasts guide staffing and resource planning, allowing the hospital to prepare for peak periods or unexpected surges in patient volume. Forecasting supports proactive decision-making, helping organizations adapt to anticipated changes without sacrificing quality of service.

Implementing Simulation for Process Improvement

Implementing simulation for process improvement allows industrial engineers to test changes, optimize resources, and enhance efficiency in a controlled, risk-free environment. By using simulation models, engineers can explore various scenarios, identify bottlenecks, and refine processes without disrupting actual operations. This approach provides a data-driven foundation for making decisions that improve productivity, reduce costs, and increase system reliability.

The first step in implementing simulation is **defining clear objectives** for what the process improvement should achieve. Objectives may focus on reducing cycle time, increasing throughput, improving quality, or optimizing resource allocation. For example, in a warehouse setting, the goal might be to reduce order picking time or improve storage utilization. Clear objectives guide the simulation, ensuring the model captures the data needed to evaluate potential improvements. Without specific objectives, simulations risk generating data that does not support actionable insights.

Data collection and model building follow the objective-setting phase. Engineers gather data on current processes, including cycle times, resource utilization, wait times, and production rates. In a manufacturing facility, this data might come from sensors, historical records, and direct observations. Collected data forms the basis of a simulation model that represents the actual process accurately. Engineers then select the most appropriate simulation type—discrete-event, continuous, or agent-based—based on the complexity and requirements of the system. For instance, discrete-event simulation (DES) is often used in production environments to model individual steps, while continuous simulation models are used for systems with uninterrupted flows, like chemical processing.

Once the data and model are set, engineers **validate and test the simulation** to ensure it accurately represents the real-world system. Validation compares simulation results with actual process data to check for discrepancies. For example, if a factory assembly line simulation shows a cycle time significantly different from the observed data, engineers adjust parameters or refine model logic until it aligns with real conditions. Validation is critical, as inaccurate models lead to incorrect conclusions. Testing also allows engineers to establish baseline performance metrics, providing a reference point to measure the impact of proposed improvements.

With a validated model, engineers conduct **scenario analysis** to test different changes and their effects on system performance. Scenario analysis helps answer "what-if" questions, such as, "What would happen if more machines were added?" or "How would reducing setup time affect throughput?" By testing each scenario in the simulation, engineers observe how the changes impact key metrics like cycle time, wait time, or resource utilization. For instance, a manufacturing simulation might reveal that adding an additional workstation reduces bottlenecks at a critical stage, boosting throughput without requiring major structural changes. Scenario analysis gives engineers a clear view of potential improvements, helping identify high-impact adjustments that support the process objectives.

Bottleneck identification and elimination are often primary goals of simulation-driven process improvement. Simulation models allow engineers to track each step's performance and visualize flow throughout the entire process. In a packaging facility, for example, a simulation might show that the labeling stage consistently causes delays, creating a bottleneck that affects downstream tasks. By analyzing these bottlenecks, engineers can test solutions like adjusting task sequences, reallocating labor, or introducing automation. Addressing bottlenecks improves system flow, reducing idle time and enabling smoother operations.

Resource allocation optimization is another valuable application of simulation. In many operations, resources like labor, equipment, and materials need to be allocated in ways that maximize efficiency. For instance, in a hospital emergency room, simulation helps determine the optimal number of nurses and physicians required to handle peak hours without overstaffing during quieter periods. By adjusting resource levels in the simulation, engineers identify the staffing patterns that balance cost with service level. Simulation models support dynamic resource allocation, making operations more responsive and cost-effective.

Sensitivity analysis in simulations tests how changes in input variables impact outcomes, providing insights into process stability and flexibility. For example, in a production facility, sensitivity analysis might involve varying demand levels or machine reliability rates to see how these factors affect throughput. Engineers can then implement safeguards or design contingency plans based on these insights, preparing the process to handle variability without compromising performance. Sensitivity analysis makes processes more resilient by highlighting critical factors that need close monitoring.

Implementing improvements based on simulation findings involves applying the most effective changes from the scenarios tested. Engineers gradually introduce these changes, monitoring actual results to verify alignment with simulation predictions. In a warehousing scenario, for instance, if simulations indicate that reorganizing item locations will reduce pick times, engineers reorganize items accordingly and track picking performance post-implementation. This phased approach allows adjustments if real-world results differ from simulation projections, refining the process over time.

After implementation, **ongoing monitoring and model updates** ensure that process improvements remain effective as conditions change. Real-time monitoring systems track key metrics, feeding data back into the simulation model to maintain accuracy. For instance, in a logistics operation, data on order volumes and delivery times can be continuously integrated into the simulation, updating the model to reflect current demand and performance. This feedback loop ensures that the simulation model remains a valuable decision-making tool, allowing engineers to make adjustments as new data becomes available or as business needs evolve.

CHAPTER 16: SUSTAINABILITY AND GREEN ENGINEERING

Principles of Sustainable Industrial Engineering

Sustainable industrial engineering focuses on designing processes that minimize environmental impact, conserve resources, and support long-term ecological balance. Industrial engineers incorporate sustainability into every stage of production, from raw material sourcing to end-of-life disposal, creating systems that prioritize efficiency, resource conservation, and waste reduction.

Resource efficiency is a core principle, emphasizing the need to use materials and energy effectively. This approach begins with selecting resources that are renewable, recycled, or sustainably sourced. For instance, using recycled aluminum instead of raw aluminum significantly reduces energy consumption. Engineers assess the resource footprint of each material and choose alternatives that lower the environmental burden. Efficiency also involves reducing water, energy, and material use at every production stage, optimizing resources while maintaining product quality.

Energy management is another critical area, as industrial processes consume large amounts of energy. By implementing energy-efficient technologies, engineers reduce greenhouse gas emissions and lower operational costs. Methods like installing energy-efficient motors, using LED lighting, and automating equipment shutdowns during idle times save energy without disrupting production. Renewable energy sources, like solar panels or wind turbines, are increasingly integrated into industrial facilities to power operations sustainably. Monitoring energy usage through sensors and analytics allows real-time adjustments to conserve energy, creating a flexible, efficient energy management strategy.

Waste reduction is essential in sustainable industrial engineering, where the goal is to minimize waste generated at each process stage. Engineers use techniques like lean manufacturing to reduce excess production and inventory, both of which contribute to waste. Process optimization focuses on eliminating non-value-adding steps, ensuring materials are used efficiently. In production, waste materials can often be recycled or reused; for example, metal shavings from machining can be collected, melted, and reformed. Engineers analyze each material flow to determine where recycling or reusing materials is feasible, creating a closed-loop system that minimizes landfill contributions.

Life cycle assessment (LCA) evaluates the environmental impact of a product from raw material extraction to disposal. Engineers use LCA to identify stages where improvements can reduce the carbon footprint, energy use, or pollution. For example, if LCA reveals that raw material extraction has a high environmental cost, engineers may explore alternative sources or materials. LCA helps industrial engineers make data-driven decisions that enhance sustainability, ensuring that products have a minimal environmental impact over their entire lifespan.

Pollution prevention strategies focus on reducing emissions and harmful byproducts. Engineers design processes that limit pollutant release, whether it's reducing volatile organic compounds (VOCs) in manufacturing or using filters to capture particulate emissions. In chemical processes, pollution prevention might involve substituting toxic chemicals with safer alternatives. Engineers also work with regulatory standards to ensure emissions are within safe limits, often implementing real-time monitoring systems that alert operators to deviations. Preventing pollution at the source reduces environmental harm and improves worker safety.

Eco-friendly product design integrates sustainability into the product itself. Engineers select materials that are biodegradable, recyclable, or less toxic, making products easier to dispose of responsibly. For example, in packaging design, engineers may use compostable materials instead of plastic. Eco-design considers factors like ease of disassembly, allowing components to be separated for recycling. The design phase is critical, as product choices directly impact resource use, waste, and end-of-life disposal options.

Sustainable supply chain management ensures that every link in the production chain adheres to sustainability standards. Industrial engineers collaborate with suppliers that prioritize sustainable practices, such as ethical labor standards and reduced carbon emissions. Sustainable supply chains consider not only the environmental impact of materials but also the logistics of transporting goods efficiently. By sourcing locally or optimizing transport routes, engineers reduce fuel use and emissions. Supply chain transparency and communication with suppliers help enforce these standards, creating a comprehensive approach to sustainability.

Continuous improvement in sustainability is a dynamic process where industrial engineers regularly review and refine systems to meet evolving environmental goals. Metrics like energy use, waste reduction, and emissions levels are tracked to assess progress and identify areas for further improvement. For instance, if energy usage spikes in certain processes, engineers investigate and adjust equipment or schedules to reduce consumption. This commitment to improvement allows industrial systems to adapt to new technologies, regulations, and environmental priorities, making sustainability a core aspect of operational strategy.

Environmental Impact Assessment

Environmental Impact Assessment (EIA) evaluates the potential environmental effects of a project or process before it begins, allowing industrial engineers to identify, mitigate, and monitor any adverse impacts. EIAs cover a broad range of areas, including air and water quality, soil contamination, waste generation, biodiversity, and community health. By conducting EIAs early, engineers develop strategies to minimize ecological disruption and ensure compliance with environmental regulations.

Baseline studies are essential in EIA, establishing a clear picture of environmental conditions before a project starts. Engineers collect data on current air and water quality, soil composition, vegetation, and wildlife in the project area. For example, before expanding a manufacturing facility, baseline studies document local water resources and air quality levels to assess potential impacts once production increases. This data serves as a benchmark, helping engineers gauge any environmental changes caused by the project.

Impact prediction involves modeling how different project activities affect the environment. Engineers use simulation tools and historical data to predict emissions, noise levels, and waste production across all project phases. For instance, in a mining project, engineers predict potential soil erosion and runoff patterns due to excavation. By understanding how each phase affects the environment, engineers make proactive adjustments to reduce harmful impacts.

Mitigation measures are designed to counteract identified impacts, minimizing the environmental footprint of the project. In industrial construction, for example, mitigation might include installing dust filters to reduce particulate emissions, designing waste treatment systems, or implementing erosion control methods to protect nearby water bodies. Engineers also consider long-term strategies, such as replanting trees or restoring disturbed areas once the project concludes. Mitigation measures are essential in EIA, as they directly address the specific risks posed by each project component.

Public consultation allows communities to provide input on the project's potential impacts. Public feedback often highlights environmental concerns, such as water quality or noise, that might not be immediately apparent to project planners. Engaging stakeholders early builds trust and ensures that engineers incorporate local knowledge into the EIA. For example, in an infrastructure project near residential areas, public consultations may reveal preferences for certain noise control measures, which engineers can integrate into their design.

Monitoring and reporting ensure that the project adheres to EIA recommendations over time. Engineers implement continuous monitoring systems to track environmental indicators, such as emission levels or groundwater quality, ensuring that they remain within acceptable limits. If monitoring reveals deviations, engineers can take corrective actions, such as adjusting operations or updating equipment. Regular reporting provides transparency to regulatory authorities and stakeholders, demonstrating the project's commitment to minimizing environmental impacts. Through ongoing monitoring, EIA becomes a living process, allowing projects to adapt to new information and maintain sustainability throughout their lifespan.

Energy and Resource Efficiency Strategies

Energy and resource efficiency strategies aim to reduce consumption, lower emissions, and minimize waste, supporting sustainable industrial engineering practices. Engineers develop strategies to maximize output using the least amount of energy, materials, and water, ultimately lowering environmental impact and operational costs.

Energy-efficient technologies are foundational in improving energy use across industrial processes. High-efficiency motors, LED lighting, and energy-optimized HVAC systems reduce electricity consumption without compromising performance. For instance, in a manufacturing plant, replacing standard motors with variable-speed drives optimizes energy usage based on workload, preventing unnecessary power consumption during low-demand periods. Automating energy use, such as turning off equipment during idle times, ensures that resources are only consumed when necessary, saving energy and reducing emissions.

Process optimization refines workflows to reduce energy needs and material waste. Engineers analyze each step to identify and eliminate inefficiencies, focusing on actions that provide value. In an assembly line, for example, streamlining machine setup times reduces idle time and lowers energy use across shifts. Batch processing and just-in-time production also cut down on waste, as they reduce overproduction and limit storage needs. By optimizing each step, engineers align energy and resource use with actual demand, minimizing excess and improving overall efficiency.

Waste heat recovery captures and repurposes heat from high-temperature processes, reducing reliance on external energy sources. In metal production, where furnaces reach high temperatures, waste heat recovery systems capture exhaust heat to preheat materials or generate steam for other plant operations. This approach reduces the total energy required by the system, cutting fuel costs and emissions. Waste heat recovery is particularly effective in energy-intensive industries, transforming otherwise lost energy into a valuable resource.

Water efficiency measures focus on reducing usage and reusing water within operations. For example, in a chemical plant, engineers implement closed-loop water systems that treat and recirculate water used for cooling or cleaning, minimizing freshwater intake. Installing low-flow fixtures and repairing leaks in piping systems also conserves water, reducing strain on local resources. Water efficiency not only conserves a vital resource but also reduces wastewater treatment requirements, lowering operational costs and environmental impacts.

Material substitution and recycling focus on replacing resource-intensive materials with sustainable alternatives and maximizing the reuse of waste. For instance, in product design, engineers may choose recyclable materials like aluminum or biodegradable plastics over non-renewable options. Manufacturing plants can reuse waste materials, such as scrap metal, within the production process or repurpose them for other uses, lowering raw material demand. Substituting less energy-intensive materials also reduces carbon emissions, making products more sustainable from production to disposal.

Smart energy management systems enable real-time monitoring and control of energy usage. These systems use sensors and analytics to track energy consumption across the facility, providing engineers with detailed insights. For example, a factory using smart meters can pinpoint areas with high energy demand, such as specific machines or departments. This data informs energy-saving measures, like scheduling heavy power use during off-peak hours, which cuts costs and reduces grid strain. Smart energy management allows for continuous adjustments, ensuring efficient energy use aligned with daily fluctuations.

Designing for Environmental Compliance and Resilience

Designing for environmental compliance and resilience ensures that industrial processes adhere to environmental regulations while maintaining the capacity to withstand disruptions. Engineers integrate these goals into every stage of production, from facility design to waste management, creating systems that minimize environmental impact and adapt to changing conditions.

Regulatory compliance is essential for environmental design, as non-compliance can lead to costly fines and reputational damage. Engineers ensure that processes meet emissions standards, waste disposal guidelines, and water quality requirements specific to their industry. For example, a manufacturing plant might install air filtration systems to capture pollutants before they're released, meeting air quality standards. Compliance is often monitored through real-time data collection and reporting systems that track emissions or water discharge, ensuring that levels remain within legal limits.

Pollution control systems are a critical component in compliant design. Wastewater treatment plants, dust collectors, and scrubbers are examples of systems that control pollutant levels. In chemical processing, engineers may design closed-loop systems that capture volatile organic compounds (VOCs) during production, preventing release into the atmosphere. Additionally, using non-toxic or less-harmful materials reduces the risk of violating environmental guidelines and minimizes the impact on surrounding ecosystems.

Resilience to climate risks is increasingly important as extreme weather events become more common. Engineers design facilities that can withstand flooding, temperature fluctuations, or power disruptions. For example, flood-resistant design might include elevated equipment and watertight doors, protecting operations in flood-prone areas. Backup power systems and energy storage solutions are also critical, allowing operations to continue during grid failures. By preparing for environmental risks, companies reduce potential downtime, ensuring that production remains stable despite external challenges.

Sustainable material selection supports both compliance and resilience. Engineers choose materials with low environmental impact and durability, such as recycled metals or sustainably sourced wood, which reduce resource consumption and withstand environmental stressors. For example, in construction, using corrosion-resistant steel not only extends the lifespan of structures but also reduces maintenance needs and waste over time.

Lifecycle resilience planning involves designing products and systems to adapt over time. Engineers anticipate future regulatory changes and evolving environmental standards, choosing designs that can be easily modified. For instance, modular machinery setups allow companies to add pollution control upgrades without overhauling entire systems. This adaptability supports long-term compliance and minimizes the need for extensive redesigns as regulations become more stringent.

Circular Economy Principles in Industrial Design

Circular economy principles in industrial design shift focus from a traditional "take-make-dispose" model to one that emphasizes resource reuse, waste reduction, and product longevity. This approach aligns with sustainable engineering by designing systems that maximize resource efficiency, reduce environmental impact, and extend the lifecycle of materials and products.

Resource recovery is a core principle, focusing on reclaiming materials at the end of a product's life. Instead of discarding used products, engineers design them for disassembly, allowing parts to be harvested and reused in new products. For example, in electronics, modular designs make it easy to replace components or reclaim valuable metals like copper and gold. By designing with disassembly in mind, industrial engineers reduce waste and lower the demand for virgin materials, which decreases the environmental footprint associated with mining and raw material extraction.

Product longevity is a key focus in circular design, encouraging engineers to develop durable, repairable products that can be reused or upgraded rather than replaced. For instance, in the furniture industry, engineers may design modular pieces where individual components like upholstery or legs can be replaced without discarding the entire item. Extending product lifespan reduces waste, conserves resources, and provides cost savings for consumers, making sustainable practices attractive on both environmental and economic fronts.

Design for refurbishment and remanufacturing enables products to be restored to like-new condition through minimal repairs or upgrades. Industrial engineers often design machinery or equipment with easily replaceable parts, allowing companies to refurbish and resell products instead of scrapping them. In the automotive industry, remanufacturing parts like engines or transmissions extends the life of components and reduces demand for new raw materials. By embedding refurbishment capabilities, engineers create products that align with circular economy goals, supporting multiple lifecycles and minimizing waste.

Materials selection in circular design prioritizes recyclability and biodegradability. Engineers choose materials that can be easily reprocessed or safely return to the environment. For instance, in packaging, using biodegradable materials allows products to break down naturally after use. Alternatively, selecting materials that maintain quality through recycling, like certain plastics or metals, supports closed-loop systems where materials are continuously reused. Engineers assess the recyclability of each material, ensuring that products contribute to a circular flow rather than linear disposal.

Industrial symbiosis in the circular economy involves sharing resources and by-products between industries. For example, in a symbiotic setup, the waste heat from a steel plant might be used to power a nearby greenhouse, or wood by-products from lumber processing could be repurposed as biofuel. By treating one industry's waste as another's resource, engineers reduce overall waste and create systems where resources circulate across industries, benefiting both operations and the environment.

End-of-life management strategies ensure that products are responsibly processed when they reach the end of their useful life. Engineers consider the end-of-life phase in the initial design, making it easy to recycle, repurpose, or biodegrade components. For instance, product labeling with recycling instructions or designing products with standardized screws and fasteners simplifies disassembly, supporting efficient material recovery. End-of-life management minimizes landfill contributions and ensures that products contribute positively to a circular system.

CHAPTER 17: OTHER TIDBITS

How to Become an Industrial Engineer

Becoming an industrial engineer involves a combination of education, skills development, and hands-on experience to prepare for a career focused on optimizing systems and improving processes. Industrial engineers work in various sectors, including manufacturing, healthcare, logistics, and technology, applying principles of efficiency, productivity, and quality management.

1. Obtain a Bachelor's Degree in Industrial Engineering or a Related Field

The first step to becoming an industrial engineer is to earn a bachelor's degree, typically in industrial engineering or a closely related field like mechanical engineering, systems engineering, or operations research. Industrial engineering programs cover essential topics such as process optimization, quality control, supply chain management, ergonomics, and systems engineering. During your degree, you will learn to use tools like discrete-event simulation, statistical analysis, and project management software, all of which are essential for a career in industrial engineering.

Some programs also offer specializations or elective courses in areas like manufacturing, healthcare, and logistics, which can help tailor your education to specific industries. Additionally, many engineering programs include hands-on projects or internships, giving you valuable real-world experience and a practical understanding of engineering principles.

2. Develop Technical Skills and Software Proficiency

Industrial engineers rely on a variety of technical skills to analyze and improve processes. Becoming proficient in relevant software tools is essential. Familiarize yourself with **statistical software** like Minitab or R for data analysis, **process modeling tools** like Arena or Simul8 for simulation, and **project management software** like Microsoft Project or Asana for planning and tracking progress. Knowledge of CAD software is also beneficial if you're involved in facility layout or manufacturing design.

Lean Six Sigma techniques are commonly used in industrial engineering to eliminate waste and improve quality. Learning the principles of Lean and Six Sigma, and even obtaining certifications like Yellow Belt, Green Belt, or Black Belt, will strengthen your skills in process improvement.

3. Gain Hands-On Experience Through Internships or Co-ops

Internships or cooperative education (co-op) experiences provide practical exposure to industrial engineering tasks and reinforce classroom knowledge. Working with established industrial engineers in sectors like manufacturing, supply chain, or healthcare exposes you to real-world challenges and teaches you how to apply engineering principles in practice. Many engineering programs have partnerships with companies, making it easier to secure internships that provide hands-on experience with tasks like process mapping, workflow optimization, and data analysis.

During an internship, you might work on improving production efficiency, designing facility layouts, or developing cost-saving measures. These experiences are valuable, as they not only build your skills but also help you determine which aspects of industrial engineering you find most interesting.

4. Build Analytical and Problem-Solving Skills

A core part of industrial engineering is the ability to analyze processes, identify inefficiencies, and design solutions. Industrial engineers must be comfortable working with data and using statistical methods to draw insights from it. Problem-solving skills are essential, as industrial engineers often encounter complex issues requiring innovative solutions. Case studies, hands-on projects, and real-world applications during your degree will hone these skills.

Developing your analytical abilities also involves learning to think critically and creatively. Familiarize yourself with problem-solving frameworks like root cause analysis, failure mode and effects analysis (FMEA), and value stream mapping. These methods allow you to approach problems systematically, ensuring thorough analysis before making recommendations.

5. Pursue Certification and Advanced Education (Optional)

While not always required, certifications can set you apart in the field. For example, the **Certified Manufacturing Engineer (CMfgE)** or **Certified Supply Chain Professional (CSCP)** certifications demonstrate specialized expertise that can enhance your qualifications. Additionally, Lean Six Sigma certifications, as mentioned, are valuable for industrial engineers looking to specialize in quality management and process improvement.

If you're interested in research or leadership roles, consider pursuing a master's degree in industrial engineering or a related field like engineering management. A graduate degree often delves deeper into specialized topics like systems optimization, advanced analytics, and decision-making under uncertainty, preparing you for roles that require a higher level of expertise.

6. Seek Employment in Industrial Engineering

After completing your education and gaining experience, start looking for entry-level positions. Job titles such as industrial engineer, process engineer, quality engineer, and operations analyst are common starting points. Industrial engineers work in diverse sectors like manufacturing, healthcare, logistics, and technology, so consider industries that align with your interests.

Networking is useful at this stage. Joining professional organizations like the **Institute of Industrial and Systems Engineers (IISE)** gives you access to job boards, networking events, and industry news. Networking with experienced professionals can provide insights into different career paths and potential job openings.

7. Embrace Continuous Learning and Professional Development

Industrial engineering is a field that continuously evolves, driven by advancements in technology and best practices. Staying current with trends like Industry 4.0, artificial intelligence, and sustainable engineering is essential. Attend workshops, webinars, and industry conferences to expand your knowledge and skills. Many engineers also benefit from mentoring relationships, where they can learn from experienced professionals who provide guidance on career growth and navigating complex challenges.

Terms and Definitions

- **Automation**: The use of technology to perform tasks without human intervention, often to increase efficiency and reduce errors.
- **Batch Production**: A manufacturing process where products are made in groups or batches rather than in a continuous stream.
- **Bottleneck**: A stage in a process that reduces the overall capacity or speed due to limited throughput.
- **Capacity Planning**: The process of determining the production capacity needed to meet changing demands.
- **Cellular Manufacturing**: A lean manufacturing approach where similar products are produced in small, self-contained production units.
- **Continuous Improvement**: Ongoing efforts to enhance products, services, or processes to improve efficiency and quality.
- **Control Chart**: A statistical tool used to monitor whether a process is in control by plotting data points against control limits.
- **Cycle Time**: The time required to complete one cycle of an operation or process.
- **Discrete-Event Simulation (DES)**: A type of simulation where events occur at distinct points in time, often used for modeling complex processes.
- **Ergonomics**: The study of designing work environments to optimize human well-being and performance.
- **Facility Layout**: The arrangement of equipment, work areas, and flow of materials within a production facility.

- **Failure Mode and Effects Analysis (FMEA)**: A method for identifying potential failures in a process and their effects to improve reliability.
- **First-In, First-Out (FIFO)**: An inventory method where the oldest items are used or sold first.
- **Fishbone Diagram**: A tool used for root cause analysis, also known as an Ishikawa diagram, to identify potential causes of a problem.
- **Gantt Chart**: A visual project management tool that outlines tasks and timelines to track progress.
- **Heijunka**: A lean production method for leveling production by reducing the impact of variations in demand.
- **Inventory Control**: The process of managing inventory levels to reduce costs and meet customer demand.
- **Just-in-Time (JIT)**: A production strategy that reduces inventory by delivering materials as they are needed in the production process.
- **Kaizen**: A Japanese term meaning "continuous improvement" through small, incremental changes.
- **Kanban**: A visual scheduling tool in lean manufacturing that controls production flow based on actual demand.
- **Lead Time**: The total time taken from the start of a process to its completion, including delays and wait times.
- **Lean Manufacturing**: A production methodology focused on minimizing waste and maximizing efficiency.
- **Line Balancing**: The process of optimizing workloads across workstations in an assembly line to eliminate bottlenecks.
- **Material Handling**: The movement, storage, control, and protection of materials throughout a manufacturing process.
- **Monte Carlo Simulation**: A method that uses random sampling to simulate and analyze the probability of different outcomes.
- **Motion Study**: The analysis of worker movements to eliminate unnecessary actions and improve efficiency.
- **Operations Research (OR)**: The use of advanced analytical methods to make better decisions and solve complex problems.
- **Optimization**: The process of making a system as effective or functional as possible.
- **Pareto Principle**: Also known as the 80/20 rule, it states that 80% of effects come from 20% of causes.
- **Predictive Maintenance**: The use of data and analytics to predict when equipment maintenance is needed before failure occurs.
- **Process Improvement**: Techniques used to make a process more efficient, effective, and adaptable.
- **Process Mapping**: The act of creating a visual representation of the workflow within a process.
- **Productivity**: A measure of output per unit of input, often used to evaluate the efficiency of a process.
- **Quality Control (QC)**: The process of inspecting and testing products to ensure they meet required standards.
- **Queueing Theory**: The study of waiting lines, useful for analyzing service processes and reducing wait times.

- **Regression Analysis**: A statistical method to determine the relationship between variables and make predictions.
- **Resource Allocation**: The process of assigning available resources in the most efficient way to achieve specific objectives.
- **Safety Stock**: Extra inventory held to prevent stockouts due to fluctuations in demand or supply.
- **SCADA (Supervisory Control and Data Acquisition)**: A system for monitoring and controlling industrial processes.
- **Service Level**: A metric that measures the percentage of customer demand that is met without stockouts.
- **Six Sigma**: A data-driven approach for eliminating defects and improving quality in processes.
- **SMART Goals**: Objectives that are Specific, Measurable, Achievable, Relevant, and Time-bound.
- **Standard Operating Procedure (SOP)**: A set of step-by-step instructions to carry out routine operations.
- **Statistical Process Control (SPC)**: The use of statistical methods to monitor and control a process.
- **Supply Chain Management (SCM)**: The coordination of production, shipment, and delivery of goods across suppliers, manufacturers, and customers.
- **Systems Thinking**: An approach to understanding a complex system by examining the relationships between its parts.
- **Takt Time**: The rate at which a product needs to be completed to meet customer demand.
- **Theory of Constraints (TOC)**: A management philosophy that focuses on identifying and improving bottlenecks in a process.
- **Time Study**: The process of analyzing and measuring the time it takes to perform specific tasks to improve productivity.
- **Total Productive Maintenance (TPM)**: A maintenance strategy aimed at increasing equipment effectiveness through regular maintenance and employee involvement.
- **Total Quality Management (TQM)**: An organization-wide approach focused on continuous quality improvement.
- **Value Stream Mapping (VSM)**: A lean tool used to document, analyze, and improve the flow of materials and information.
- **Visual Management**: The use of visual tools like signs, charts, and indicators to communicate key information in a work environment.
- **Waste Minimization**: Strategies used to reduce the amount of waste produced in a process.
- **Workflow Analysis**: The examination of work processes to improve efficiency and reduce bottlenecks.
- **Work Measurement**: The process of determining the time required to complete a job under specific conditions.
- **Work-in-Progress (WIP)**: Items that are in production but not yet completed.
- **Yield**: The amount of product produced compared to the amount of material input, often used as a measure of process efficiency.
- **Zero Defects**: A quality management concept that aims for producing products with no defects or errors.

AFTERWORD

Thank you for taking the time to journey through *Industrial Engineering Step by Step*. I hope this book has given you not only a deeper understanding of industrial engineering but also practical insights that you can apply, whether in your career or daily life. Industrial engineering may seem like a technical field on the surface, but at its heart, it's about finding solutions to real-world challenges and making systems work better for everyone involved.

Throughout the chapters, we covered a lot of ground: from foundational concepts to specific tools and techniques, from designing efficient workflows to improving ergonomics and sustainability. Industrial engineering has a wide reach, impacting industries from manufacturing to healthcare, technology, and even service-oriented sectors. My aim was to break down each topic into manageable, understandable steps, showing how each area contributes to the larger picture of operational excellence. I hope you feel equipped with a toolkit of concepts that you can pull from when you encounter a problem that needs a fresh perspective.

As you move forward, remember that industrial engineering is a continuously evolving field. New technologies, methods, and challenges are constantly emerging, especially as data analytics, AI, and sustainability reshape the way we approach engineering. Staying curious, adaptable, and committed to lifelong learning will be invaluable as you build your career or explore new opportunities within this field.

If you're an aspiring industrial engineer, I hope this book has sparked your enthusiasm and given you the confidence to pursue this path. For those already working in the field, perhaps this book has provided a fresh look at familiar concepts or introduced new ones you can apply in your work. And if you're simply exploring industrial engineering out of curiosity, I hope you now have a new appreciation for the ways this field impacts our everyday lives, often in unseen but significant ways.

Industrial engineering is about more than numbers, charts, and process maps; it's about improving the systems we interact with daily and making life a little more efficient and enjoyable for all. It's a field that constantly challenges us to think critically, solve problems creatively, and work toward meaningful change.

Thank you once again for reading. May you find success and fulfillment in applying what you've learned, whether you're designing the next breakthrough process or simply finding ways to make your corner of the world run a bit more smoothly. Here's to a future filled with creativity, efficiency, and innovation!

Made in the USA
Las Vegas, NV
11 May 2025